Cobalt Blues

Portrait of Leonard Grimmett 1949 [1]

Peter R. Almond

Cobalt Blues

The Story of Leonard Grimmett,
the Man Behind the First Cobalt-60 Unit
in the United States

Peter R. Almond
Houston
USA

ISBN 978-1-4614-4923-2   ISBN 978-1-4614-4924-9 (eBook)
DOI 10.1007/978-1-4614-4924-9
Springer New York Heidelberg Dordrecht London

Library of Congress Control Number: 2012945098

© Springer Science+Business Media New York 2013
This work is subject to copyright. All rights are reserved by the Publisher, whether the whole or part of the material is concerned, specifically the rights of translation, reprinting, reuse of illustrations, recitation, broadcasting, reproduction on microfilms or in any other physical way, and transmission or information storage and retrieval, electronic adaptation, computer software, or by similar or dissimilar methodology now known or hereafter developed. Exempted from this legal reservation are brief excerpts in connection with reviews or scholarly analysis or material supplied specifically for the purpose of being entered and executed on a computer system, for exclusive use by the purchaser of the work. Duplication of this publication or parts thereof is permitted only under the provisions of the Copyright Law of the Publisher's location, in its current version, and permission for use must always be obtained from Springer. Permissions for use may be obtained through RightsLink at the Copyright Clearance Center. Violations are liable to prosecution under the respective Copyright Law.
The use of general descriptive names, registered names, trademarks, service marks, etc. in this publication does not imply, even in the absence of a specific statement, that such names are exempt from the relevant protective laws and regulations and therefore free for general use.
While the advice and information in this book are believed to be true and accurate at the date of publication, neither the authors nor the editors nor the publisher can accept any legal responsibility for any errors or omissions that may be made. The publisher makes no warranty, express or implied, with respect to the material contained herein.

Printed on acid-free paper

Springer is part of Springer Science+Business Media (www.springer.com)

To my wife June who for the last ten years has indulged my passion for Grimmett and the cobalt unit

Foreword

Leonard George Grimmett was an Englishman who came to Houston in 1949, sight unseen, to start a physics department at a brand new institution. Unfortunately, he died suddenly of a heart attack 28 months after arriving in Houston. I had no intention of writing his biography and was in fact researching the life of another Englishman who had come to Houston 37 years earlier, also sight unseen, to start a physics department in a new institution. His name was Harold Albert Wilson who came to Houston in 1912 as chairman of the physics department at the Rice Institute, the year Rice admitted its first students. Wilson was one of a group of extraordinary young men who had studied physics at the Cavendish Laboratory at Cambridge University at the turn of the century under J. J. Thomson. Four of the group would receive Nobel Prizes, J. J. Thomson (1906), Ernest Rutherford (1908), C. T. R. Wilson (1927), and Owen Richardson (1928), and yet it was said that H. A. Wilson was the best in the group. Given that fact I was intrigued as to why Wilson would come to a small town on the edge of civilization to an institution that had not even started, especially when universities like Princeton and Columbia were interested in him. I was researching the answer to these questions for articles to be published by the Rice Historical Society when I came across some of Grimmett's early letters. Wilson's sister had married Owen Richardson and although Wilson kept few, if any, of his personal papers, Owen Richardson kept everything and filed them away, including letters to his wife and mother-in-law who in later life lived with them. Being a recipient of the Nobel Prize in Physics, his papers were of interest to science historians and after his death they were acquired by the University of Texas. While reviewing Harold Wilson's letters in these files in Austin, I was surprised to see the name of Leonard George Grimmett in the catalogue. I knew who Grimmett was but did not know of the connection between Grimmett and Richardson. The files contained letters from Grimmett to Richardson spanning nearly 20 years from 1926 to 1943. They started when Grimmett was a young undergraduate at King's College in London, where Richardson was chairman of the physics department, seeking to do research for Richardson, continued while Grimmett was a research student under Richardson and finally when he was a professional medical physicist. Although the correspondence ended while

Grimmett was still working in London I wondered if he knew, when he decided to come to Houston, that Richardson's brother-in-law, Harold Wilson, was head of the physics department at the Rice Institute in the same town; did that play any part in his decision to come and did they ever meet?

As a result of the articles published by the Rice Historical Society I learned that Grimmett's secretary[1] was still alive and living in Houston. When Grimmett died unexpectedly she had had the responsibility of collecting his personal affects; not knowing or being told what to do with them she put them in a suitcase where they had been kept in her attic for over 50 years until I contacted her. The contents of the suitcase began to answer the above questions and helped explain the close connection between the departments of physics at Rice University and the M. D. Anderson Cancer Center. It also opened up a whole new world about Leonard Grimmett, his life, and his quest to improve upon the tele-radium treatment units by replacing the radium with a more suitable lower costing artificial radioactive isotope.

A word about institutional names: many of the institutions mentioned in this account have undergone various name changes over the years. When recounting specific events I have used the names of the institutions as they were known by at the time of the event. When the institutions are referred to in a more general context I have used the names they are known by today. For example, Rice University was the Rice Institute until 1960 when the name change took place. For most of this account therefore it will be called the Rice Institute. The University of Texas M. D. Anderson Cancer Center (MDACC) is the current designation for what was called, when Grimmett arrived in Houston, the M. D. Anderson Hospital for Cancer Research of the University of Texas. This in most cases has been shortened to M. D. Anderson Hospital (MDAH), the name it is still referred to by the local population.

The Oak Ridge Institute for Nuclear Studies (ORINS) changed its name to Oak Ridge Associated Universities (ORAU) in 1966. Since this was long after the events in this book ORINS will be used throughout.

For the physicist the meaning of the designation 'medical physicist' or its derivatives, for example, 'hospital physicist' or 'physicists in medicine', has not changed over the years. The same is not true for the clinicians. The general term for physicians using radiation is 'radiologist' both in imaging, where the more specific term 'diagnostic radiologist' might be used, and in therapy. However, to distinguish the physicians who treated with radiation from those who diagnose with radiation, the term radiotherapist came into use around 1950. Today the term 'radiotherapist' designates the technologists who treat patients on the machines and the MDs are now called 'radiation oncologists'. Since that term had not come into use during the time period of this book, the term 'radiotherapist' will be used in this book unless the modern term 'radiation oncologist' is more appropriate.

---

[1] Her name was Shepley, née Kocian. She died in 2007, while this book was being written but not before she met Grimmett's last surviving relative, his niece.

Exact quotes from transcripts, memos, letters, newspaper accounts, and interviews are either in quotation marks or indented.

The book has been written using only simple physics and mathematical concepts. The design and use of a cobalt-60 treatment machine depends upon some understanding of the treatment of cancer with radiation and the medical physics concepts involved and these have been outlined in Appendix A. The actual production of the radioactive cobalt-60 requires some knowledge of activation of materials in a nuclear reactor. Since this is the area in which the M. D. Anderson Hospital machine ran into problems that delayed its initial use and resulted in, what is called here, the "Cobalt Blues", a short primer on production of radioactive cobalt-60 in a reactor is given in Appendix B.

Acknowledgments

This book would not have been possible without the help and support of many people and I am indebted to all of them. In particular, Lesley Brunet and Javier Gaza at the Historical Resource Center of the University of Texas M. D. Anderson Cancer Center made available to me the archives of the University of Texas M. D. Anderson Cancer Center and the Texas Medical Center and provided help and insight in locating material helpful to this project.

Mrs. Jane Dyer of England, Grimmett's niece and only surviving family member, provided much useful information. Not only about Grimmett's family but located material about his career from the National Archives at Kew in England.

Mrs. Trudy Shepley née Kocian who had been Grimmett's secretary in Houston provided invaluable information about Grimmett's time at M. D. Anderson Hospital and had kept some of his personal papers (which were later donated to the Historical Resource Center) without which this project would not have been possible.

I am indebted to a number of people who knew Grimmett when he was in Houston and who graciously spent time with me recounting their memories of him, including Jorge and Harriet Awapara and Mrs. Jasper Richardson. In particular Dr. Robert Shalek, whom Grimmett hired as a graduate student and who later became chairman of the physics department, provided many interesting insights about Grimmett and his relationship to Fletcher.

I wish to acknowledge my thanks to Dr. Thomas Fletcher and Mr. Walter Fletcher, Fletcher's sons for very helpful conversations about their father and for donating invaluable material to the Historical Resource Center at the University of Texas M. D. Anderson Cancer Center.

In addition to the Historical Resource Center at the University of Texas M. D. Anderson Cancer Center the John P. McGovern Historical Collections and Research Center at the Houston Academy of Medicine-Texas Medical Center Library also provided helpful documentation.

I also wish to express my thanks to the UNESCO archives in Paris, France, to the Welcome Library in London, England, the University Archives of the University of Illinois in Urbana, Illinois, to the Harry Ransom Center at the

University of Texas, and the Niels Bohr Library and Archives at the American Institute of Physics and their helpful staffs for helping locate additional material about Grimmett.

I am grateful to Mona Tomlison and her friends in Mississippi for providing information about the Critz family and providing files of the Starkville Daily News that greatly helped with the time-line of this story.

To my colleagues who were supportive throughout I am very grateful, including Ken Hogstrom, Jack Cunningham, and Ralph Worsnop and in particular to Roger Robinson who for many years has been interested in this subject and who willingly shared all the material he had obtained from Marshall Brucer.

My thanks to Walter Pagel and the staff of the department of Scientific Publications at MDACC, especially John McCool, who were very helpful throughout and to Scharlene Wilson whose help in putting the manuscript together was crucial.

I am forever thankful to Christopher Coughlin Physics Editor for Springer who was willing to undertake this project and for the professionals at Springer who made it possible.

Thanks to all of you and to many more who willingly gave of their time and expertise to get this project completed.

Prologue

On August 6, 1945 during World War II the United States of America exploded an atomic bomb over Hiroshima, Japan, followed 3 days later by a second bomb dropped on Nagasaki. Over 100,000 Japanese were killed and both cities were devastated. The next day, August 10, 1945 Japan surrendered and World War II ended. Almost one year later on August 1, 1946 the United States Atomic Energy Act was signed into law transferring the control of atomic energy from military to civilian hands, under the auspices of the United States Atomic Energy Commission (USAEC). It was strongly felt that atomic energy should be used to promote world peace and improve the public welfare as much as, if not more than, for nuclear weapons.

In Tennessee, The University of Tennessee joined with 14 other southern schools to form the Oak Ridge Institute of Nuclear Studies (ORINS) to take advantage of the opportunities offered by the Oak Ridge National Laboratory that had been an integral part of the atomic bomb project. On October 17, 1947 ORINS received its charter of incorporation. Early in its history medical research became an important focus and in 1948, the Atomic Energy Commission authorized ORINS to establish a clinical research program to study the use of radioactive materials in treating and diagnosing diseases and to set up a cancer research hospital. A newspaper headline of the day declared: *"Cancer Cure found in the Fiery Canyons of Death at Oak Ridge,"* referring to thyroid treatment with radioactive iodine [2].

The Manhattan project that had developed the atomic bomb was a joint effort between US, British, and Canadian scientists and a large number of British scientists had moved, during the war, from Britain, mainly to Canada but some to the United States, to aid in the effort. The British medical physicist, Leonard George Grimmett was working for the Medical Research Council (M.R.C.) in London at the time and enquiries were made about his participation but:

> He declined to assist in the atomic bomb development. 'I don't mind killing Germans in odd numbers' he said with a wry grin, in oblique explanation [3].

Grimmett was an expert in the use of radium to treat cancer and in the safe handling and measurement of radiation and radioactive materials in clinical situations. He had spent the best part of his career devising better, safer, and more efficient ways to treat cancer with radiation and he remained in England during the war. Then in 1948 while working for UNESCO in Paris he received an offer he could not refuse the, "… *post as physicist to a new 'Cancer Research Institute and Atomic Center' in The University of Texas*", [4] one of the original universities in the ORINS' consortium. Thus was set in motion the events that would lead Grimmett to Houston, Texas and to be the first person to publish, in 1950, the design of a cobalt-60 radiation therapy unit for the treatment of cancer. For the next 25 years cobalt-60 units would be the mainstay of cancer radiation therapy, treating millions of patients worldwide. Grimmett, however, would not live to see the completion of his work. This is his story.

Logo of the U.S. Atomic Energy Commission [5]

Contents

1   M.D. Anderson Cancer Center, 1941–1949 1

2   The Journey, January 29 to February 7, 1949 9

3   Early Life and Education, London, 1903–1929 11

4   Medical Physicist Part I, London, 1929–1944 15

5   The Unknown Years and UNESCO, Paris, 1944–1948 29

6   Replacing Radium, 1937–1949 35

7   The Arrival, Houston, February 1949 41

8   The Cobalt Unit, 1949–1954 51

9   Medical Physicist Part II, Houston, 1949–1951 73

10 Cobalt-60 and the Notebook 97

11 Cobalt-60 in Perspective 103

12 Erratum to: Cobalt Blues E1

Epilogue: Grimmett the Man 121

Appendix A: Principles of Radiotherapy 125

Appendix B: Principles of Reactor Production of Cobalt-60 131

Appendix C: Grimmett's Suggested References on Cobalt-60 135

References 139

Index 151

Abbreviations

A.E.C.	Atomic Energy Commission
AECL	Atomic Energy of Canada Limited
BIR	British Institute of Radiology
BOAC	British Overseas Airways Company
C.C.I.R.U.R.S.I.	International Consultative Committee for Radio Communications of the International Union of Radio-Science
D.S.I.R.	Department of Scientific and Industrial Research
DBM	Division of Biology and Medicine
G.E.	General Electric
HPA	Hospital Physicists' Association
JAMA	Journal of the American Medical Association
MIT	Massachusetts Institute of Technology
MDACC	MD Anderson Cancer Center
MDAH	MD Anderson Hospital
MRC	Medical Research Council
NRX	National Research Experimental Reactor
ORINS	Oak Ridge Institute for Nuclear Studies
UNESCO	The United Nations Educational, Scientific and Cultural Organization
UNO	United Nations Organization
UNSCEAR	United Nations Scientific Committee on the Effects of Atomic Radiation

Chapter 1
M.D. Anderson Cancer Center, 1941–1949

In 1942, the University of Texas Board of Regents announced that the Texas State Cancer Hospital, which had been created the previous year by an act of the State Legislature, would be located in Houston. As a temporary site for the hospital, the M.D. Anderson Foundation had acquired the "Oaks", the estate of the late Captain Baker, from the Rice Institute, and the hospital was to be named the M.D. Anderson Hospital for Cancer Research of the University of Texas (Fig. 1.1).

The "Oaks" was located approximately three miles southwest of downtown Houston. Captain James A. Baker had been a prominent attorney in the law firm of Baker, Botts and Baker and represented many wealthy citizens, one of whom was William Marsh Rice, founder of Rice Institute. After Rice's death in 1900, a suspicious Captain Baker alerted authorities to the possibility of foul play. Because of his efforts, investigators discovered that an associate had in fact murdered Rice and had produced a false will. Captain Baker made sure that the perpetrators were prosecuted and convicted and that Mr. Rice's valid will bequeathing a large amount of money for the establishment of an institute of higher learning in Houston was probated. Under the guidance of Captain Baker, the William Marsh Rice Institute in Houston was established. When he died in August 1941, one month after the state legislature had approved the formation of the new hospital, he left the "Oaks" to the Rice Institute.

The announcement of the location, in temporary quarters, for the new hospital along with the appointment of Dr. Ernst W. Bertner as temporary director was published in the Journal of the American Medical Association (JAMA) on September 26, 1942 (Fig. 1.2).

In 1937, Monroe Dunaway Anderson had created the M.D. Anderson Foundation. He was a founder of Anderson Clayton and Co, which at the beginning of the twentieth century was the foremost cotton-merchandising concern in the world. Mr. Anderson moved to Houston in 1907 as his firm's representative, and he became a great benefactor of the city. Among the benevolent and charitable purposes that he had outlined for the foundation were "the promotion of health, science, education, and advancement and diffusion of knowledge and

Fig. 1.1 Baker Estate main residence building [6]

understanding among the people" [8]. Although he died in 1939, the trustees of the foundation envisioned, in keeping with the above purposes, the development of a great medical center in Houston, and in 1943, they purchased a little over 134 acres for such a center adjacent to Herman Hospital, one of Houston's largest hospitals, located about six miles south of downtown. They were not, however, the first to consider the site for a medical center. A real estate developer, Will Hogg, son of a former governor of Texas, was the first person to envision a medical center on the property and had purchased this site some years previously. When the medical center did not materialize, he sold the property to the City of Houston for use as a park, and the sale of the land to the M.D. Anderson Foundation had to be approved by the people of Houston in the fall election of 1943. The Texas Medical Center was charted under the laws of the State of Texas in 1945, and the following year, the M.D. Anderson Hospital for Cancer Research of the University of Texas was the first institution to be approved for inclusion in the center. Although the groundbreaking ceremonies for the hospital were held in 1950, the cornerstone was not laid until 1953 due to the shortage of construction material brought on by the Korean War. In the meantime, the hospital continued to operate in the facilities at the Baker estate. The new facility was dedicated in 1954, and the institution moved from the Oaks estate to the Medical Center. Rooms with thick concrete walls, designed by Leonard Grimmett, had been constructed in the

Fig. 1.2 JAMA announcement of the establishment of the Texas State Cancer Hospital Project and Research Laboratories [7]

September 26, 1942

TEXAS STATE CANCER HOSPITAL PROJECT UNDER WAY

The M. D. Anderson Foundation, Houston, recently purchased the 6 acre estate of the late Captain James A. Baker for use as temporary quarters of the Texas State Cancer Hospital and Research Laboratories. The property will be donated to the University of Texas as temporary quarters for the hospital and laboratories until a permanent plant can be built. The last legislature appropriated $500,000 for the project and the Anderson Foundation agreed to donate a site and to give $500,000 in addition, according to the state medical journal. On August 1 the board of regents of the University of Texas appointed Dr. Ernst W. Bertner, Houston as temporary director of the hospital and laboratories. He was president of the Texas State Medical Association in 1938.

Journal of the American Medical Association

basement of the hospital to house the institution's radiation therapy equipment including a cobalt-60 treatment machine. Although Grimmett was present at the ground breaking ceremonies in December of 1950, he died the following year and did not see the hospital completed.

The first permanent Director of the new hospital, Dr. R. Lee Clark, was appointed in 1946, and he set about hiring a staff and getting the institution off the ground. As a result of the atomic bomb research during World War II, radioactive isotopes were becoming available, and in order to promote the peaceful uses of radioisotopes, the Atomic Energy Commission and the Federal government were awarding research and construction grants for the application of radioactive isotopes in medicine. Dr. Clark decided that if he could hire the right people, this would be an area in which the new hospital could make a significant contribution, and he set about recruiting a suitable staff and petitioning construction funds for this purpose from the state and federal governments.

In 1947, Dr. Clark met Dr. Gilbert Fletcher who had trained in diagnostic and therapeutic radiology in Europe and New York Hospital. Gilbert Hungerford Fletcher was born in Paris, France, in 1911 to an American father and a French mother. His early education was in Paris, but the family moved to Brussels around 1929, and he attended the University of Louvain (B.A. Civil Engineering 1932) and the University of Brussels (M.A. mathematics 1935 and a medical degree in July 1941). In medical school, he studied diagnostic and therapeutic radiology where he was introduced to the use of radium sources and radium teletherapy treatments for cancer. Although he had been born in Paris, Fletcher was an American citizen because of his father. In 1941 World War II had been going on for two years, but the USA was still neutral. Belgium was an occupied country, and Fletcher realized that it was best for him to leave as soon as possible, before the USA entered the war. He returned to France and then made his way across the Pyrenees into Portugal and to Lisbon where he obtained passage to New York City. He entered a residency program at Cornell medical school in radiology at the New York Hospital where he met Mary Walker Critz, from Starkville, Mississippi, who was doing a fellowship in pediatrics; they were married in 1943.

Fletcher had always intended to become a radiotherapist because of, as he said, his "… training in engineering and mathematics prior to my medical training" [9]. But after his residency in radiology in 1945, he was drafted into the U.S. army where he practiced diagnostic radiology for two years in Pittsburgh. Fletcher had decided that after the army, he would go to Europe for extra training in radiotherapy, visiting several major cancer centers, but he also needed to find a cancer hospital that would hire him as a radiotherapist on his return. When Fletcher was discharged from the army in the spring of 1947, Fletcher was 36 years old, married and had a young son, and he and the family went to Starkville, Mississippi, to visit his wife's parents. The Starkville News April 25, 1947, reported their arrival:

> Dr. and Mrs. Fletcher of Pittsburgh, Pennsylvania, arrived Wednesday for a visit in the home of Mrs. Fletcher's parents, Mr. And Mrs. Harry Critz [10].

The Critz's family doctor in Starkville was Dr. J. F. "Fetty" Eckford, who had come to know Dr. Clark when Dr. Clark had spent two years as a surgeon, prior to World War II, in Jackson, Mississippi, helping to establish a state medical school. Fletcher had read the announcement in JAMA about the new cancer hospital in Houston and that Clark had been appointed the Director. When Dr. Eckford heard of Fletcher's interest in working in a cancer hospital, he wrote a letter of introduction to Dr. Clark. The Fletchers and Mrs. Critz were planning to drive to Edinburgh, Texas, to visit with Mrs. Fletcher's sister, which was duly reported in the Starkville News for May 2, 1947:

> Dr. and Mrs. Gilbert Fletcher and Mrs. Fletcher's mother, Mrs. Harry Critz are in Edinburgh Texas, for a visit with Mr. and Mrs. W.H. Utz [11].

They therefore planned to stop in Houston, on their way, to meet with Clark. Clark remembered the meeting:

> So one day in walked this fellow in a captain's uniform with a heavy French accent. He said, 'I'm Gilbert Fletcher and I want to work in a cancer hospital. But I need to go back to Europe. I haven't been over there for a good while and I need to go back and see what they're doing in therapy' [12].

One of Clark's assets was his ability to quickly sum up people and recognize their potential, and after about an hour's conversation, Clark offered to help Fletcher with his plans in exchange for a comprehensive report on his findings. "We'll appoint you right now as our first traveling Anderson fellow," [12] he said.

Fletcher remembered the event slightly differently:

Shortly after my arrival in England, I wrote him about the wonderful things I was learning. Following this letter, Dr. Clark gave me an appointment as a traveling fellow with the charge to bring back all possible information [9].

Three weeks later, on May 24, 1947, Fletcher arrived in Southampton, England, aboard the United States Lines' the "Marine Marlin". He went immediately to the Royal Cancer Hospital (Free) in Fulham Road, London. Here, he met Dr. Manuel Lederman whose area of interest was the radiation treatment for head and neck cancer. As a result of this meeting, Fletcher and Lederman became close personal friends. At the Cancer Hospital, Dr. Lederman had the use of a 10-g radium teletherapy unit that had been designed by Leonard Grimmett to treat the head and neck cancer patients. Dr. Lederman was also interested in the development and use of applicators to treat cervical cancer, and at the time of Fletcher's arrival, Lederman and the physicist Lamerton had just read a paper to the British Institute of Radiology on the dose estimation and distribution in the radium treatment for cervical cancer. Head and neck cancer and cervical cancer would become the main focus of Fletcher's career and the fields upon which he hoped to base his reputation. Fletcher would also have had interaction with the chairman of hospital's physics department, V.W. Mayneord. Mayneord had recently returned from one year in Canada at the Chalk River Project (part of the atomic bomb program), where he had been sent by the British Government to determine the potential for nuclear physics in medicine. He and A.J. Cipriani (from Chalk River) had just sent a paper to the Canadian Journal of Research on the absorption of gamma rays from cobalt-60 in which they specifically mentioned the possibility of cobalt-60 replacing radium in certain therapeutic applications [13]. It is not known, however, whether this was discussed with Fletcher at that time. Mayneord also gave a series of lectures at the British Institute of Radiology on the potential of the new artificial radionuclides for medical use. Fletcher immediately wrote Clark about the new things he was learning and sent him the material he had so far gathered, which probably influenced Clark to come up with the concept of establishing an "Atomic Center" at the new hospital to investigate the use of radioactive isotopes for medical use.

A few months later, Dr. Fletcher was at the Radiumhemet in Stockholm where again he would have seen treatments carried out on a radium teletherapy unit, this one designed by the Swedish physicist Dr. Rolf Sievert. The two units were very similar since the British unit followed closely the Swedish design. Both Sievert in Sweden and Mayneord in London would have stressed the need for a qualified physicist in any department undertaking radiotherapy. Fletcher would also have been aware of this, having known the medical physicists Edith Quimby in New York and Professor Piccard in Brussels. His problem was how to find a qualified physicist willing to come to a new cancer hospital in Houston on his recommendation alone, assuming he himself was offered a job there.

Fortunately, while in Europe, he heard, probably from both Mayneord and Sievert, that a senior medical physicist might be available for such a proposition: Leonard George Grimmett, the man who had designed the radium teletherapy units in Great Britain. At the end of the war, Grimmett had left medical physics and was at that time employed by The United Nations Educational, Scientific and Cultural Organization (UNESCO) in Paris, but was interested in getting back into medical physics. Fletcher probably knew of Grimmett through his publications in the radiology journals prior to the war. Although living in Paris, Grimmett also maintained his home in London and made frequent trips back there. Fletcher also visited Paris during this trip, but it is unlikely that they met there since Grimmett was traveling extensively in Mexico and USA for UNESCO while Fletcher was in Europe. It is much more probable that they met, around New Year's Day 1948, in London, when Grimmett was back there for the holidays. Fletcher's name does not appear in Grimmett's diary for the year 1947, but it is the first name that appears in the address section of his 1948 diary. Fletcher's address is given as Starkville, Mississippi, USA, which was Mary Fletcher's hometown, where she was living and working while Gilbert Fletcher was in Europe. In his 1947 diary, Grimmett wrote twelve pages of notes on the winter meeting of the Hospital Physicist's Association (HPA) held at the British Institute of Radiology in London on January 2, 1948. The meeting included a report on the status of radiotherapy and medical physics in USA, and Grimmett took care to record the information that was presented. It seems highly likely that Fletcher might have stayed over in London a few extra days during the time of this meeting to gauge the possibility of hiring a physicist from England. Fletcher recalled that, "Part of my charge as a traveling fellow was to contact possible recruits for M D Anderson Hospital, and in that capacity, I met in London an English physicist, L. G. Grimmett" [9]. If Fletcher had just talked to Grimmett about joining him in Houston, it would account for Grimmett's careful notes on the United States and for including Fletcher's American address at the beginning of his 1948 diary [14].

Fletcher sailed out of Southampton for USA on the American President Lines' "Marine Falcon" on January 7, 1948.

The Starkville (Mississippi) News, the weekly newspaper for Starkville, ran the following note in the Friday February 20, 1948 edition:

> Dr. and Mrs. Gilbert Fletcher arrived early this month for a visit with Mrs. Fletcher's parents, Mr. and Mrs. Harry Critz. They will go in a few days to Houston, Texas, where they will both practice medicine [15].

When Fletcher returned to the United States, he was appointed head of the radiology department at M.D. Anderson Hospital, taking up his appointment in the spring of 1948. With Dr. Clark's concurrence, Fletcher began to recruit Grimmett to move to Houston to help establish an atomic energy radiology center in the permanent facility to be built in the Texas Medical Center.

At this time, the concept of replacing the radium in one of Grimmett's pneumatically operated tele-radium units with radioactive cobalt-60 began to take shape. It is likely that this evolved during the correspondence between Grimmett, Fletcher and Clark concerning Grimmett's appointment at M.D. Anderson Hospital.

In April of 1948, David E. Lilienthal, Chairman of the U.S. Atomic Energy Commission, had announced the discovery and production of inexpensive radioactive cobalt that might eventually become a substitute for radium in the treatment for cancer [16]. A Field Notice (4-30-48) from the American Cancer Society followed, which advised:

> While it is uncertain when radioactive cobalt will become widely available for clinical use, you are advised to defer, if possible the purchase of radium for the use in cancer clinics... [17].

Subsequent to that notice, Fletcher wrote a memo to Clark in 1948:

> Radioactive cobalt could develop into a substitute for radium if it becomes much cheaper....
>
> When radioactive cobalt will be cheap enough, it will make the possibility of a radioactive cobalt bomb within reasonable cost and will make a very interesting project, both physically and clinically...
>
> This ground work (sic) requires the presence of an excellent work shop (sic) manned by good instrument makers...
>
> It is feared that, until the really experienced physicists in that field are with us and adequate equipment and an adequate research fund is available, no real valuable work can be done [18].

This sounds very much like part of Fletcher's arguments to get Grimmett appointed and probably indicates some of Grimmett's requirements in order for him to accept the position. Grimmett was almost fanatical about the need for a well-equipped workshop and a qualified instrument maker (machinist). Fletcher recalled that:

> His (Grimmett's) first item of business was to develop a physics shop, which he thought was indispensable to the design of radiotherapy equipment. I was in agreement since I, myself, had witnessed what can be done in such a shop [9].

There was also interest in the medical applications of other radioactive isotopes, and Clark began planning for an atomic and radiology research center, for which he sought state, federal and private funds.

Before coming to M.D. Anderson Hospital, Clark had been chief of surgery at Randolph Field in San Antonio at the army's School of Aviation Medicine where he had come to know Dr. Shields Warren. Clark sought federal grants through the A.E.C. writing on February 5, 1949 (two days before Grimmett arrived in Houston), to Dr. Shields Warren for support, who, at the time, was chairman of the Division of Biology and Medicine (DBM) of the A.E.C:

> As a cancer research institution we have felt that we have a particular interest in working with the isotopes. The program that we have envisioned has been one of research in biophysics and the clinical application of radio-active (sic) materials of any kind...
>
> Dr. Leonard G. Grimmett... would be particularly suitable in working out the clinical applications of Cobalt 60, and we would like to construct a replica of his pneumatic teleradium unit as a telecobalt unit.
>
> Do you think there is a possibility of receiving help of this nature from the Atomic Energy Commission? [19]

At the same time, Clark sent a copy of this letter to the local congressman, Albert Thomas:

> You were so helpful to us in securing the Public Health grant that I was hopeful you might have some suggestions regarding this program, he wrote [20].

The M.D. Anderson Foundation pledged $1,350,000 of matching funds and $700,000 was allocated from federal funds, both subject to state funding, but the Texas Senate cut the funds from the appropriations bill. This led to headlines in the local newspapers.

The Houston Post reported April 1, 1949:

> Cancer Work in Peril City May Lose Atomic Center [21].

On the same day, the paper ran an editorial on "Atomic Cancer Treatment," stating:

> If the legislator grants the needed funds, it will make possible a program which will bring to Texas for the first time a completely designed and equipped facility for the use of radioactive materials [22].

Other newspapers took up the cause, and the protest lasted through April. Fortunately, in May, the senate finance committee restored the recommended appropriations and brought the bill before the special session of the fifty-first Legislature, which passed the bill appropriating $1,350,000 for construction of the atomic energy radiology facilities. On March 1, 1950, Governor Allen Shivers signed the bill, "one of the best investments the State ever made," he said [23].

Dr. Grimmett's journey from his boyhood home in north London to Houston and the amazing contribution he made to the fledgling institution are the subjects of this book. Even today, 60 years later, his influence on the physics department at the M.D. Anderson Cancer Center is still apparent and is one of the main reasons why the department is considered among the best in the world. In 1976, Dr. Clark would say of him:

> Leonard Grimmett was absolutely the man for the job. He was about forty-five years old. He died when he was forty-nine, but he had outlined the total physics department of Anderson hospital as it is today. It was fantastic—the accumulated knowledge in that man's mind and the way he could put it down [24].

Chapter 2
The Journey, January 29 to February 7, 1949

Grimmett left London for Glasgow by night train on Saturday evening January 29, 1949. The sleepers were fully booked, and he had to sit up all night and only occasionally dozed off. Arriving in Glasgow on Sunday morning, he took a bus to Prestwick airport close to the town of Troon on the west coast of Scotland where, he wrote his wife, was a famous golf course.[1]

Transatlantic air travel was still developing in the late 1940s after the war, and the flights were along the old military routes. Prestwick airport was the main airport used during the war for military planes flying between North America and Great Britain. He boarded a BOAC (British Overseas Airways Company) flight to New York, which took off at 3:00 p.m. on Sunday January 30. But there were strong head winds in the North Atlantic, and soon after take-off, the plane was diverted to Keflavik airport in Iceland, arriving 4 h later. Keflavick airport had been built, in 1943, by the United States as a military air base for the war effort and in 1949 was operated by American civilian companies. The air terminal was situated in the middle of the base, and civilian air travelers had to enter military check points to reach their flights. Grimmett's passport has a stamp by the US Immigration and Naturalization Service of the Justice department that simply says, "Admitted 1/30/49." It is the last dated entry related to travel, in his passport. Apparently, this was all he needed when he eventually entered the continental USA. Finally getting clearance, the plane took off again for Gander in Newfoundland. The temperature in Gander, when they arrived, was −10° F, and he reported that it was the coldest he had ever been. The plane had trouble with its tail plane or as he called it, the rudder, and it took two hours to fix.

Once again they took off and this time made it to Montreal by Monday afternoon, where BOAC put the passengers up in the Laurentien Hotel. The weather over North America was now so bad that all the US airports in the northeast had been closed, and the passengers had to take the train to New York. Leaving

---

[1] He was referring to the Royal Troon Golf Club. The British Open Championship had been played there in 1923 and was scheduled to be the site of the 1950 British Open Championship the following year.

Montreal at 8:30 p.m. Monday evening, they arrived in New York at 8:00 a.m. Tuesday morning February 1.

Wednesday February 2, he took the train to Boston to meet with John Trump, Head of the High Voltage Research Laboratory at MIT and the technical director of the High-Voltage Engineering Corporation, builders of Van de Graff accelerators. Grimmett was very interested in the comparison between the Van de Graaff accelerator and the betatron for radiotherapy purposes. He had built a Van de Graaff accelerator for the Medical Research Unit at Hammersmith Hospital in London several years previously but had never had the opportunity to completely test it. Even though the train journey to Boston took 5 h each way, he thought the trip worthwhile and told his wife what a lovely time he had there. He left New York the next afternoon, Thursday February 3, for Washington D.C. where he stayed at the historic Willard Hotel.

Friday February 4, he spent in Washington, renewing acquaintances with several friends and colleagues from UNESCO. He left for Houston on Saturday February 5. He had now been traveling for a week and had not yet reached his final destination. He arrived in Houston Monday afternoon February 7, 1949, and went straight to the M.D. Anderson Hospital housed in the "Oaks", the old Baker estate, at 2310 Baldwin Street. What he found greatly shocked him!

He wrote his wife the next day:

> And the hospital! Well, words fail me! Its (*sic*) true that they told me it was in sheds, but I wasn't prepared for anything so primitive [25].

Chapter 3
Early Life and Education, London, 1903–1929

Leonard George Grimmett was born in Tollington Park, North London, on September 12, 1903. His father was a paperhanger and an interior painter who had been medically discharged from the British Army in 1917 during World War I. Grimmett was the eldest of three boys, Leonard (born 1903), Horace (born 1904) and Reuben (born 1906), and his father insisted that they all take music lessons, but on different instruments, Leonard's instrument was the piano. After attending elementary school, he was awarded a Junior County Scholarship to The Holloway County Secondary School when he was 12 years old. Here, he distinguished himself in French, chemistry and physics. He must have also been an outstanding music student because in 1921, at age 18, upon completion of his schooling, the Royal Mail Steam Packet Co employed him as a musician. He played the piano on their ships to South America, and he stayed in Brazil and Argentina for 2 years, earning his living playing music. He returned to England in 1923 and entered the University of London's King's College. In 1926, he was awarded an honors degree (B.Sc.) in physics with a second in pure mathematics. The chairman of the department was Professor Owen Richardson, who would receive the 1928 Nobel Prize in Physics. Lecturers in the department at the time were Henry Flint, Bernard Worsnop[1] and Edward Appleton (who would receive the 1947 Noble Prize in Physics).

Upon graduation and on the recommendation of Professor Appleton, Grimmett petitioned Richardson to become a graduate student at King's. In September of 1926, he wrote Richardson:

> I…want something to do in the research lab at King's College. I should be very pleased to know if you can find something for me to start on; I am anxious to commence as soon as possible [26].

---

[1] Bernard Worsnop's son Ralph Worsnop was a member of the physics department at M.D. Anderson Hospital in the early 1960s.

He was accepted as a graduate student, and in 1927, he applied for a grant from the Department of Scientific and Industrial Research (The D.S.I.R.) for support, and in July, he wrote Richardson telling him he had the grant:

> I have pleasure in informing you that the Department of Scientific and Industrial Research have awarded me a grant of £140 for one year. I thank you for your recommendation [27].

This was equivalent to $8,500 in 2010 dollars and was hardly sufficient for Grimmett to live off. He went on to explain to Richardson that he was having difficulty getting the chemicals he needed for his research, and it would be 3 weeks before he would receive supplies from Germany. He concluded with:

> I wonder if I might take a short holiday during my enforced idleness [27].

Idleness it seemed was not a characteristic that Grimmett tolerated very well, although later in his life, he would be called "bone-idle" by others.

In November 1928, Grimmett sent Richardson his annual report for 1927–1928 that he was required to send to the D.S.I.R., noting that it was late. A year later, he wrote Richardson a short note thanking him for a check and sending him, "Heartfelt Christmas greetings from myself and wife" [28]. This is the first indication that he had married.

In fact, he had married Norah Anastasia Gill the year before on September 25, 1927, in the Roman Catholic Church of St. Joan in Highbury, London, when he was 24 years old and she was 23 years old. She was a ballet dancer who had trained under Pavlova,[2] and her father worked for the London County Council as an employment inspector.

His research at King's College under Richardson was on the emission of electrons under the influence of chemical action and was published in the Proceedings of the Royal Society in 1930 [29].

In 1929, Grimmett left King's College without receiving his Ph.D and went to work at the Westminster Hospital in London with Dr. Henry Flint. He described this period of his life in a letter to Richardson dated May 18, 1932:

> 1929. The progress of science robbed me of my living as a musician, so I went to Westminster Hospital [30].

Just what he meant by this is not clear. It is known that he supported himself while at the University by playing the piano in restaurants and theaters. For any student going to university in England at that time, it was highly unusual for them to support themselves by working at the same time. Playing the piano at night must

---

[2] Anna Pavlova was born in St Petersburg in 1881. She studied in the Imperial Ballet School attached to the Mariinsky Theatre. She made her debut at 17, and by 1906, she had become the Mariinsky's principal ballerina. In 1907, she made her first foreign tour, and in 1908, on her second, joined Diaghilev's Ballets Russes. In 1912, she purchased Ivy House in Hampstead, England, where she established her own school of dance. She made her last appearance in St. Petersburg in 1913 and spent the rest of her career almost constantly on tour, bringing ballet to millions for the first time through the drawing power of her legendary name.

3   Early Life and Education, London, 1903–1929

have been unique and stressful, and Richardson was completely discouraging, believing that a degree could not be taken in this eccentric fashion. The year 1929 saw the end of the silent movie era, and Grimmett probably could no longer get work playing for silent movies. John Read recalls the following incident:

> As we walked down Regent Street in 1934 we passed seedy, out-at-elbow violinists playing in the gutter. He was filled with distress-'I know this kind of man; I have played with them. They are good musicians displaced by the Talkies' [31].

Pay for a junior medical physicist at that time could not have been very much, and his stipend as a graduate student had been very little, certainly not as good as playing the piano, and he complained to Richardson in the letter,

> Unfortunately I have been unable to send the requisite fee for the Ph.D. exam. My anxiety for a long time past has been just to go on living [30].

It would be another 10 years before he submitted the fee for his degree.

Chapter 4
Medical Physicist Part I, London, 1929–1944

Westminster Hospital Annex.

When Grimmett left King's College, he became Assistant Physicist at the Westminster Hospital Annex, working under Professor H.T. Flint, and they developed one of the earliest hospital physics departments in the United Kingdom. Professor Henry Thomas Flint was an extraordinary man. He was a successful academic physicist, medical physicist and radiotherapist (i.e. radiation oncologist in today's terminology) and was someone that Grimmett greatly admired.

At the Westminster Hospital, Grimmett became a pioneer in the development of radium beam units participating in the building of a 4 g radium teletherapy treatment machine (or radium bomb) which was described in joint papers with Professor Flint [32, 33]. The development of treatment units would become a life-long pursuit of his.

He also became interested in radiobiological problems working with Dr. F. G. Spear of the Strangeway's Laboratory at Cambridge on the influence of gamma-ray intensity on the inhibition of mitosis in tissue cultures. This resulted in two publications, the first in 1933. The 4 g of radium at the Westminster Hospital were used to irradiate tissue cultures and the response was determined by the effect upon cell division [34]. Grimmett had initially hoped to measure the intensity of the radiation but there was too much leakage in his ionization chambers and the intensities had to be calculated. This work earned them a runner-up award to the 1934 Garton Prize and Gold Medal. This award, instituted by the British Empire Cancer Campaign in 1929, was for an essay on the nature, causes, prevention, and treatment of cancer. The subject for the 1934 assay was, "The Biological Effects and Mode of Action of radiations upon Malignant and other Cells." The Times of London reported on April 11, 1934 that:

> It was decided that the award of £500 and the Gold Medal should be awarded to Dr. H. A. Colwell…As one of the other essays was of such high merit the grand Council decided that a second award of £100 should be made to its authors, Dr. F. G. Spear in association

with Dr. R. G. Canti, Mr. L. G. Grimmett, Dr. B. Holmes, Miss S·F. Cox and Dr. W. H. Love. [35]

If the prize were split evenly among the authors Grimmett would have received a little less than £17, but even that amount would have been a welcome bonus for him.

The second paper, two years later in 1935, gave the results of follow-up measurement using Sievert type condenser chambers. By this time however, only 1 g of radium was available but he was able to show agreement in the intensity distribution between the measured and calculated values [36].

Grimmett's tenure at the Westminster was quite successful but threatened to be short-lived. The use of radium in treatment machines appeared to be useful, especially in treating head and neck cancers. But radium was scarce and very expensive. Westminster hospital had tried two designs for treatment units, one based on a treatment machine in Paris which had not proved too useful and the other of Flint and Grimmett's design. But the radium was owned by the Radium Commission and in 1932 they decided to divide the 4 g of radium into four separate 1 g amounts. The Westminster hospital would retain 1 g and the rest was distributed to three other hospitals. One gram of radium made treatments excessively long and in most cases impracticable. The occasion of Grimmett's 1932 letter to Richardson was to ask him for a reference for a new post.

> I thank you for your offer to give me a reference for the post of Assistant Physicist at the Cancer Hospital (Free), Fulham....
> I have published two papers dealing with γ-ray measurements.
> At the present moment I am still at Westminster Hospital which threatens to close down the dept. for lack of funds, so I am applying for the post at the Cancer hospital, where a research worker is required to carry on some experimental work dealing with the spectroscopy of hard X rays and general application of radium to biological work.
> P.S. Will you kindly send the testimonial to me, as I have to make 4 copies. [30]

Two days later on May 20 1932 Grimmett wrote Richardson thanking him for the testimonials,

> and for the kind remarks contained therein. [37]

Westminster Hospital did not close down the department and Grimmett did not get the position at the Cancer Hospital. A better offer came along.

The two papers that he mentioned resulted from his research on the use of radon seeds. They were on a direct-reading γ-ray electroscope that used a Lindemann electroscope and the control circuit for the electroscope that made use of the Townsend balance method [38, 39]. The electroscope required a very high value resistor and he developed methods of making these in the range of 10^{10}–10^{11} ohm.

This and the work with Spear caught the eye of Professor McLennan, Chairman of the Executive Committee of the Radium Beam Therapy Research Unit. In 1934 Grimmett was persuaded to join the Unit that was formed "to investigate the treatment of cancer by the use of radiation from large quantities of radium at a distance," (large quantities meant 5 g or more) and Grimmett was one of the few physicists in the field at that time with experience with multiple grams of radium.

Professor Cunningham McLennan D.Sc. F.R.S. had retired from the chairmanship of the physics department at Toronto University at age 65 in 1932 and he had moved to England to live. While in Canada he had served on the Canadian Radium Commission and when he retired he became a member of a committee in London to consider and report on the value of tele-radium therapy. McLennan was concerned when the 4 g radium source at Westminster Hospital was divided up believing that single larger amounts of radium were needed to investigate any possible benefit to tele-radium therapy.

He proposed a Radium Beam Therapy Research unit at the Radium Institute in Portland Place in cooperation with the Royal Cancer Hospital (Free) in Fulham Road, under the auspices of the Medical Research Council (MRC) [40]. Radium was valued at £10 per milligram at that time (approximately $750 in 2010 dollars) and he put together a formal agreement between Union Miniére du Haut Katanga of Belgium (first part), the Governing Board of the Radium Beam Research Council (second part) and The Cancer Hospital (Free) (third part). Union Miniére was to make a loan of 5 g of radium with the Cancer Hospital buying 1 g, a half-gram for £5000 immediately and the other half-gram for another £5000, twelve months later. This was a total investment of £10,000 or $750,000 in 2010 dollars. This was a large amount of money but they got the use of radium costing five times that amount. A five member Executive Committee was set up with three members from the Hospital and two from the Radium Institute with McLennan as Executive Secretary. All publications and advertising had to have prior approval of the Hospital House Committee (which controlled staff and hospital policy). McLennan consulted with W.V. Mayneord the physicist at the Cancer hospital and the hospital physics department was to be available to the Radium Beam Therapy Research.

The agreement took from September 27, 1934 to February 28, 1935 to complete and the loan agreement was finalized by 14 August, 1935.

Constance A.P. Wood a young radiotherapist was appointed the Resident Clinical Research Officer on loan from the Fulham Rd. Cancer Hospital and Leonard Grimmett was appointed the physicist.

Just when Grimmett was appointed and took up his responsibilities is not known. The 1950–1951 M.D. Anderson Hospital Annual Report, which recorded Dr. Grimmett's death, states that he spent a year in Stockholm at the Radiumhemmet [41]. No record has been found supporting this. The visitors book at the Radiumhemmet records that he visited the institution on October 20, 1933 [42]. In the 1934–1937 Report on the Radium Beam Therapy Research Grimmett wrote that he spent several weeks at the Radiumhemmet and also that at the end of 1933 he visited Professor Claude Regaud at the Foundation Curie in Paris [43].

In September 1932 Professor McLennan and others had also visited Rolf Sievert at the Radiumhemmet in Stockholm and saw the 3 g radium unit designed by Sievert. In July of that year they had also visited Professor Claude Regaud in Paris. It therefore seems likely that Grimmett left the Westminster hospital sometime in mid-1933 and traveled to Stockholm and Paris at the suggestion of Professor McLennan, and when he returned he took up his position with the Radium Beam Therapy Research Unit in London at the time the agreements concerning the Unit were being finalized.

Radium Beam Therapy Research

As the work of the unit got underway in 1934 the best treatment technique to be adopted was carefully considered from all angles with the result that the Stockholm apparatus and methods were chosen.

When Grimmett visited Stockholm Sievert had just published a paper on his design for a radium treatment unit that was quite innovative [44]. Similarities between that unit and one that Grimmett would design can clearly be seen [45]. This was also one year after Rolf Sievert had been award the doctoral degree for his work on ionization chambers and the Sievert condenser ionization chamber was to greatly influence Grimmett's approach to making radiation measurements. His first use of such chambers was probably the follow-up measurements that he did for Dr. Spear on the tissue culture experiments.

It was while he was at the Radium Beam Therapy Research unit that he learned to get along with people with strong personalities. Read reported that Grimmett:

...was kind, gentle, always soft spoken and quite imperturbable. He rejoiced that McLennan was a dynamic, forceful personality. "He will uphold the claims of physics with the Medicals" he said, ignoring the likelihood that he might himself be the first to suffer from such forceful character. Nor was he; McLennan might hustle others but Grimmett went his own way. [3]

This was a characteristic that Grimmett would have to call upon the rest of his life. Grimmett might have been able to handle McLennan but he would eventually run head-on into Dr. Constance Wood director of the radiotherapy unit and he would not come off so well in that encounter. And later still he would clash with Fletcher's strong personality but by that time he was prepared to deal with it!

Unfortunately Professor McLennan died suddenly in October 1935 while traveling on the Continent and Sir Edward Mellanby was appointed to replace him.

John Read, who was a colleague of Grimmett's at Westminster Hospital wrote Grimmett's obituary for the British Journal of Radiology and in it noted that Grimmett made a great contribution to the MRC,

...especially in the design of such equipment as the 10 gm. beam unit with pneumatic transfer of the radium and the perspex man. [3]

The perspex or celluloid man, as it was also called, was one of the first anthropomorphic phantoms made [46]. It was a life-size model of the human body with celluloid plates and air spaces such that the overall density was unity. Grimmett knew that for radiation protection considerations the total energy absorbed by a person was required rather than the dose to any particular point in the body, and the model incorporated a large ionization chamber to measure the total body dose for any particular set of conditions.

The treatment unit, or radium bomb, consisted of a storage-safe for the radium and the unit (treatment head) with the radium being transferred between the two by a pneumatic system [45]. This allowed more time to be spent adjusting and positioning the unit with respect to the patient when the radium was in the safe and no radiation was present in the treatment room. Upon completion of the patient set up the personnel could leave the room and the radium was pneumatically moved into the treatment unit. The main part of the treatment head was made of tungsten alloy and designed originally for about 5 g of radium. The design of this unit and its radiation shielding properties were so good, however, that eventually it was safely loaded with 10 g of radium [47] (Fig. 4.1).

Grimmett was involved in all aspects of the design and use of the unit. He studied the radiation shielding properties of the tungsten alloy, he helped develop a patient head stabilizing device to minimize patient movement during treatment, and he was concerned with and measured the dose of the personnel running the unit [48, 49]. He also worked on the design of Sievert-type condenser ionization chambers and in particular on "air-wall" materials for their construction. In order to make a suitable "air-wall" material that was wave-length (energy) independent, Grimmett and his co-workers formed a composite mixture of bakelite and graphite with a small percentage of titanium or vanadium oxide added. This material could be conveniently molded under pressure and resulted in chambers with excellent mechanical and electrical properties [50].

In 1937 he attended the fifth International Radiology Congress in Chicago crossing the Atlantic on the famed ocean liner, the Queen Mary, the largest liner built to that date [51]. Before going to Chicago he went to Madison and visited with Donald Kerst who was experimenting with the concept of a betatron. He sent his wife a post card on September 13 of Lake Monona, incorrectly referring to the university as Madison University rather than the University of Wisconsin at Madison.

> Am just on my way back to Chicago from a glorious day at Madison University. The Congress starts tomorrow. [52]

Grimmett was very interested in the betatron's possible application to radiotherapy. Dr Failla from the Memorial Hospital in New York arranged a dinner party at the Congress at which Grimmett sat next to Kerst where their discussions continued on the use of accelerators for radiotherapy.

This trip was a defining moment in Grimmett's life. It exposed him to the research going on in the United States in a number of areas that he strongly thought the Medical Research Council in Great Britain should be looking into including accelerators for the production of neutrons and artificial radioactive substances both of which he believed had a future in medicine. He also felt that Great Britain was behind in producing high voltage (in the one MeV range) x-rays.

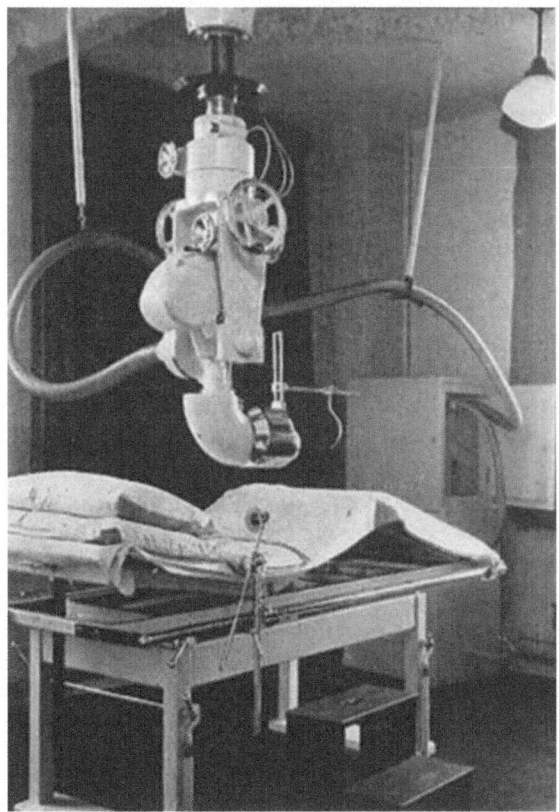

Fig. 4.1 Grimmett's 5 g unit showing the container safe and interconnecting pipe for the pneumatic transfer of the radium [47]

Upon his return to Great Britain he wrote to Sir Edward Mellanby a report on his trip urging the use of artificial radioactivity and the building of a high voltage machine to make isotopes and to produce neutrons for therapy [40]. He also wrote about this in papers to the British science journal, Nature, and included his thoughts in the Report of the Radium Beam Therapy Unit, which was published in early 1938, as a special report by His Majesty's Stationary Office [53, 54, 55] (Fig. 4.2).

Few copies of this report now exist. The copy at MDACC has this hand written note on the inside of the cover:

> Presented to the library of the M.D. Anderson Hospital for Cancer Research. My only copy: all of the printer's stocks were destroyed in the fire-raid on London Dec 1940.
> Signed: Leonard G Grimmett
> Feb 1951 [43]

Dr Wood's copy is in the Welcome Medical Library in London with her personal papers.

Part of the report, written by Dr. Constance Wood, was on the clinical work concentrating mainly on the treatment of head and neck cancer patients. The rest

Crown Copyright Reserved   Special Report Series, No. 231

Privy Council

MEDICAL RESEARCH COUNCIL

REPORT ON RADIUM BEAM THERAPY RESEARCH

1934–1937

by

Constance A. P. Wood, L. G. Grimmett,
T. A. Green and others,

under the direction of the Governing Body of Radium
Beam Therapy Research

LONDON
HIS MAJESTY'S STATIONERY OFFICE
1938

Universal Decimal Classification
615.849.7

Fig. 4.2 Title page of the 1937 Report on Radium Beam Therapy [55]

of the report was on the technical aspects of the program, written by Grimmett. No mention was made in the report, however, of the cooperative work with the Royal Cancer Hospital. The radiation treatments of the patients using the Grimmett designed tele-radium unit looked very promising and the press picked this up. In November and December 1938 the British newspapers including The Times, The Sunday Chronicle, The News Chronicle, and The Daily Express reported on the work in treating cancer with radium beams. Constance Wood and Grimmett's names were frequently mentioned as being the physician in charge of the treatments and as the physicist responsible for the design of the equipment respectively. Sir William Bragg, President of the Royal Society and Chairman of the Governing Body of the Radium Beam Therapy Research was quoted as being hopeful for success for the Radium Beam Therapy Research. The interest was so great that on Thursday December 6, 1938 Sir William Bragg went on the 9:00 o'clock B.B.C. evening news and read a very optimistic report of the work. Although Grimmett's name is not mentioned Bragg talked about his work including the "perspex" man and the pneumatic transfer radium tele-therapy unit.

This publicity however, did not sit well with the people at the Royal Cancer Hospital who felt very strongly that the agreement, requiring joint approval before any publication, had been violated. Sir William Bragg heard of some dissatisfaction with his broadcast but was assured by the B.B.C. that he had said nothing inappropriate. The House Committee at the Cancer Hospital, however, was outraged. Dr. Constance Wood defended her action by pointing out that none of the clinical results she reported on involved patients from the Cancer Hospital, since the follow-up on those patients was too short; but to no avail. She was required to resign from the Royal Cancer Hospital but continued as Head of the Radium Beam Therapy Research at the Radium Institute.

Grimmett was also involved. On January 23, 1939 he wrote Dr. Wood that he had heard rumors that, "I had incorporated research work done jointly with the Cancer hospital in our recently published report" [40] and that he had not credited the cooperation. This complaint had come from Mayneord who objected to the fact that Grimmett had not recognized his input into resolving the output of the radium units.

Grimmett had written in the report that:

The value of dosage given in the report have been 'corrected to correspond to 8.3 r/hr per milligram' of radium at 1 cm distance, filter 0.5 mm platinum. [55]

Grimmett was not happy about the dispute and attempted to resolve the conflict by telephoning and writing Mayneord about it and suggesting they publish a joint paper on the subject and he thought he had Mayneord's agreement for this. But Mayneord was still not satisfied and he wrote Grimmett on January 30, 1939 still complaining about the situation. By this time Grimmett had had enough and he replied to Mayneord on February 21, 1939. He was surprised that Mayneord was still not satisfied, he pointed out that the two radium units (one at the Radium Institute and the other at the Cancer Hospital) had differed in output by 25 %, due to differences in the construction of the two units and in the way the measuring instruments had been calibrated and that Mayneord had abandoned his calibration and accepted Grimmett's which brought the two units into agreement. He also pointed out that the 8.3 factor had been widely adopted and neither he nor Mayneord could lay claim to it. He concluded that under the circumstances he no longer whished to proceed with their joint publication. By this time, however, the agreement between the Cancer Hospital and the Radium Beam Therapy Research had been dissolved (January 30, 1939).

In early 1939 the Radium Beam Therapy Research unit started a collaborative effort with the Imperial College, London to build two high voltage electrostatic generators following closely the design of the pressure-insulated generators of Van de Graaff and Trump; one for basic physics research would remain at Imperial College, the other for medical purposes at the Radium Beam Therapy Research Unit. During the blitz in 1940 The Radium Institute in Portland Place in central London was bombed and Grimmett's workshop was damaged. In December of that year he and his physics group moved to the Imperial College of Science and Technology in South Kensington to work on the high voltage Van de Graff generator which now had priority second only to war work. Work on an investigation to compare the clinical results of treatments using radium (γ-rays) and 200 kvp X-rays that had commenced in 1939 was brought to a halt due to the war.

When World War II began in late 1939 steps were taken to make sure that the large amounts of radium that the unit had were stored so that it would be safe in case of bombing and be hidden from the Germans in the event of invasion. Some of the radium was put in boreholes up to 50 feet deep and 12 inches in diameter. Efforts were undertaken to move the Radium Beam Therapy Research out of central London and possibly out of London altogether. Cambridge, Oxford, Birmingham, Manchester all expressed interest in the Unit, along with several other London County Council hospitals, including Hamersmith Hospital. Grimmett and Constance Wood made numerous trips to all these places to investigate their suitability to house the Unit. This put them in close contact with each other and in their reports on these visits there is no hint of any animosity between them. In April of 1940 they visited Hammersmith Hospital as a possible site for the unit and this turned out to be the place to which the Radium Beam Therapy Research unit eventually moved on a permanent basis. In 1941 the Medical Research Council assumed full responsibility for radium beam work and the research team was reconstituted as the Council's Radiotherapeutic Research Unit under the direction of Dr. Wood.

The equipment from Radium Beam Therapy Research at the Radium Institute in Portland place was moved during 1941 and 1942 to Hammersmith Hospital. The Van de Graaff generator construction was also moved to Hammersmith Hospital from Imperial College.

For the first time in three years the clinical and physics' departments were together again. In addition to the Van de Graaff construction research resumed on the clinical comparison between gamma rays and X-rays [40].

In 1942 Grimmett wrote Kerst asking further information about his accelerator, the betatron. Kerst replied with details about his machine and sent copies of a paper that had just been published in the Review of Scientific Instruments. He also commented on the electrostatic accelerators and stated that he was "sure that your electrostatic machine will do well for you, and I hope that some day you can also find a betatron valuable" [56]. This was advice that Grimmett would not forget. Grimmett thanked Kerst for his letter and the reprints. On July 7 1942 he wrote Kerst:

> Congratulations on having achieved such success with this new method. It is likely to be of great use in many branches of physics.
> Your letter also encourages me to persevere with our pressure-generator, which I intend to as far as the war will allow. Things seem to get more difficult as the months go by. We all expect to be called up for war service ultimately, but until then we shall struggle along as best we can! [57]

Beside his scientific work he was also participating in several other activities. In 1943 he was involved in the founding of the Hospital Physicists' Association (HPA) in Great Britain. One of the first national medical physics groups formed. A group of British medical physicists met at the British Institute of Radiology (BIR) in London in September of 1943 and agreed that '—a body should be formed to interest itself in and discuss matters arising out of natural interests of those engaged in hospital physics' [58]. A follow-up meeting was held, again at the BIR, in January of 1944 and Grimmett was appointed as one of four ordinary members to the Executive Committee. He immediately suggested that 'the Association should consider the possibility of running a central workshop, on a non-profit basis, for the provision of certain essential instruments and research apparatus'. Unfortunately because of lack of funds and the difficulties managing such a shop would pose, it never materialized. It is likely that the ongoing war also had something to do with this decision.

Grimmett had interests beyond physics. He practiced fine calligraphy and book binding, and was a script writer [41].[1]

He was a gemologist and a worker in precious metals and grew sapphires, which he made into jewelry and had a small laboratory at home for doing this. As a sideline he went into partnership in the jewelry business with his brother Ruben. In 1948 his share of the profit from this business was £180 [59]. This would have been about one-half to one-third the annual salary he had made as a medical physicist and so would have represented a rather large sum of money to him. He also was a pilot. In the letters between Grimmett and his wife when Grimmett first arrived in Houston there is an ongoing discussion between them if she should pack his flying jacket and bring it to Houston and whether they would have time or the opportunity in Houston to fly [60].

In 1943 he decided to complete the requirements for his Ph.D. degree. He wrote Professor Richardson from Hammersmith Hospital on June 29, 1943:

---

[1] This could possibly have been scripts for silent movies that appeared on the screen to the accompaniment of the piano. Grimmentt was known to have played for the silent movies when he was at King's College.

Professor Sir Owen Richardson
Dear Sir Owen,
I wish to apply for the degree of Ph.D. in respect of the work I carried out under your supervision in 1926–1929.
Would you sign the enclosed statement for me, if you consider it correct, and also the entry form?
I am afraid I have left it rather a long time [61].

In 1929 Richardson had received the 1928 Nobel Prize in physics and had been knighted for this honor. Grimmett gives no reason why he has left it so long to apply except that what he really wanted, he said, was a D.Sc. degree and was told that getting his Ph.D. first would help, so he was now applying for that degree. It should be noted that this was what Professor Flint had done and undoubtedly this greatly influenced him. The statement that Grimmett sent to Richardson was a declaration attested by the supervising teacher that the work for the Ph.D. degree was carried out under his supervision and that the work was done completely by the student. After some minor changes Richardson signed the declaration on July 19, 1943 [62].

But when did Grimmett get his Ph.D.? In the official history of the HPA he is consistently referred to as Mr. L. G. Grimmett, whereas the other members with a Ph.D. are referred to as "Dr". The biographical data in the history, however, lists the year of his Ph.D. as 1930 [63]. The research for the degree was definitely completed in 1929 but for financial reasons it appears that Grimmett did not make formal application with King's college until 1943 [64]. It seems likely, however, that the formal paper work on his degree was not completed in time for it to be awarded in that year. The records at King's College show that the degree was not awarded until November of 1946, three and one half years after he initiated the necessary paperwork. Why the delay? 1944 and 1945 appear to be very difficult years for Grimmett [65].

He continued to work at the Radiotherapeutic Unit of the Medical Research Council at Hammersmith Hospital until 1944 but by that time he and the Director Dr. Constance Wood did not get along. In mid 1944 Dr. Wood was complaining about Grimmett's work performance, that he was not doing much and that his two assistants, Boag and Howard Flanders[2] were complaining about him. Although Grimmett was pleasant enough she considered him bone-idle. He did not show up for work until around 11 o'clock, and sometimes as late as 3:00 p.m., and left at 5:00 p.m. He claimed much of his time was taken up with secret work (war related) that Dr. Wood knew about. This whole matter went to the Secretary (Chief Executive Officer) of the Medical Research Council, Sir Edward Mellanby. On September 13, 1944 Mellanby requested that Boag come and see him and discuss the situation at the Radiotherapeutic Research Unit at Hammersmith and the

---

[2] Both Jack Boag and Paul Howard Flanders went on to distinguished careers in medical and biophysics. Jack Boag as Professor of the University of London in the Institute of Cancer and Paul Howard Flanders as Professor of Molecular Biophysics and Biochemistry at Yale University.

situation quickly developed after that. Grimmett's primary responsibility, at that time, was the completion of the electrostatic generator which he claimed could be finished in one or two months if the appropriate parts arrived on time. Boag disagreed and thought that at least twelve months were required with the whole staff working on it. Mellanby had heard from others, which Boag confirmed, that Grimmett had been anything but assiduous about working on the apparatus.

Boag also remarked on Grimmett's work hours further stating that often he did not go to the laboratory more than three times a week. Boag and Grimmett had had their disagreements and Grimmett had accused Boag of wanting to take over his job. Both Boag and Howard Flanders had offered to resign and had submitted their resignation letters in August; Howard Flanders stating that he could no longer continue professional relations with Dr. Grimmett. When asked what Grimmett did with his time Boag said he did not know. He knew that Grimmett had been working on a gamma-ray detector for G. P. Thomson[3] which he had been officially requested to do. However, for two weeks at least Grimmett, Boag said, had been making jewelry. Boag also told Mellanby that the design of the electrostatic generator was not original but a copy of an American machine that had been published and which Grimmett had seen. In Mellanby's eyes this made the problem much more serious and he determined to call in an independent expert to review the situation.

On September 15, 1944 a couple of days after Boag's visit with Sir Mellanby Grimmett went to see him on his own initiative saying he would do his best to complete the high voltage generator. Grimmett was told that it was too late for that and that his colleagues had offered their resignations. Mellanby also offered to bring in a senior physicist of standing to report on Grimmett's work during the previous two years. Now it was Grimmett's turn to say that the time was too late for that. It was then suggested that Grimmett take up a position with the Commission on Scientific and Laboratory Equipment that had previously been offered him, but he seemed not to be interested. Finally Sir Mellanby gave Grimmett to understand that he was completely dissatisfied with his work and led him to believe that his only choice was to join the Commission on Scientific and Laboratory Equipment. Grimmett blamed the situation on Dr. Wood interfering with his work.

Three days later September 18, 1944 Professor Sidney Russ was called into review the situation and determine how long it would take to complete the high voltage generator. Professor Russ was the senior medical physicist in Britain at that time having started at the Middlesex Hospital in London in 1910. He was a precise and at times an autocratic person but could be genuinely kind and understanding. He was considered a good friend to his colleagues and to science in

---

[3] G.P.Thomson was the son of J.J. Thomson they both received the Noble prize for physicis. J.J. in 1906 and G.P. in 1937; a rare father son combination. In 1930 G. P. Thomson was appointed Professor at Imperial College in the University of London. He was made Chairman of the British Committee set up to investigate the possibilities of atomic bombs. This committee reported in 1941 that a bomb was possible, and Thomson was authorized to give this report to the Americans.

medicine and an encourager of the young. No better choice could have been made to review a difficult and tense situation. His report agreed with Boag's assessment of the situation. By September 30, 1944 Mellanby could report that the situation at the unit was resolving itself. Grimmett had offered his resignation in writing and that the council would accept it. Grimmett was on loan elsewhere for six months and would not be returning to Hammersmith [65].

Chapter 5
The Unknown Years and UNESCO, Paris, 1944–1948

When he left the Hammersmith Hospital in 1944, no one is sure quite what Grimmett did next. It is likely that he went to the position with the Commission on Scientific and Laboratory Equipment, which Sir Mellanby had mentioned, and maybe this was where he was on loan for six months. The records at M.D. Anderson Hospital show that during the war, he was a consultant to the Ministry of Aircraft Production on the manufacture and use of radioactive luminous compounds, although this hardly would have been a full-time position [41].

Little is known about Grimmett during this time as indicated in his obituary, and not much more information is contained in the biographical sketch in the History of the Hospital Physicists' Association:

> It is not clear where or how Dr. Grimmett spent the five years between leaving Hammersmith and arriving at the M.D. Anderson except that he was for a time working for UNESCO. He said he had been coping with 'radiation sickness' [63].

It is not surprising that he may have suffered from radiation sickness, since it is known that the radium source in the teletherapy radium unit that Grimmett had designed would sometimes stick in the tube between the safe and treatment unit when being pneumatically transferred and Grimmett would have to free the source. He definitely discussed his radiation sickness with his colleagues when he arrived in the United States in 1949. He said that he had been assigned to construct Geiger counters (it is likely these are the radiation detectors he was building for G. P. Thomson) and that initially he could not get them to work correctly because of excessive leakage [66]. This was not resolved until he learned that the room in which he had been assigned to work was once used to process radium sources, and there was enough contamination in the room to cause his Geiger counters to continuously discharge. A new site was found to build the detectors. This must have also resulted in Grimmett receiving a large radiation dose. In any event, he did record that he had suffered from radiation sickness but apparently was recovered when he went to UNESCO since no record of his having suffered from it can be found in his personnel file there. However, his doctor in London, Dr. Snowden, advised him to get weekly blood counts when he arrived in Houston, and if the counts deteriorated, he should rest up [67]. The radiation sickness may have added to the reasons for the delay in getting his Ph.D degree.

He was, for two years (1944–1946), Secretary of the Science Commission of the Conference of Allied Ministers of Education. The Conference of Allied Ministers of Education was created in London in 1942 to forestall, by educational means, totalitarian regimes taking over countries. It was the organization that would subsequently give birth to UNESCO. He joined the Preparatory Commission of the United Nations Educational Scientific and Cultural Organization (UNESCO) on April 1, 1946. It seems highly likely therefore that the prospect of a position with UNESCO prompted him to finally get his Ph.D. He had an initial appointment for three months, at the end of which there was still work for him to do, and he wished to continue in the position. Julian S. Huxley was the first Director General of UNESCO, and he was advised about Grimmett's wishes. It is likely that Grimmett knew Huxley since he was professor of Zoology at King's College London when Grimmett was a student there. UNESCO was officially established on November 4, 1946, in Paris, and Grimmett became a "Programme Specialist-Apparatus Information Officer" in the Natural Sciences Department, working primarily with the Scientific Reconstruction Program. He resigned from UNESCO on January 18, 1949 [4].

At UNESCO, Grimmett worked in the Natural Science Division under fellow Englishman, Joseph Needham, and when Needham was absent in early 1947, Huxley asks Grimmett to act for Needham. Needham left the division in April 1948, and the staff gave him a farewell party, which seems to have been arranged by Grimmett. Grimmett kept a copy of the menu for Needham's farewell dinner, signed by most of the members of the division, and the items have a definite physics connection. For example, item No. 7 on the menu was "Bombe atomique glacée praliné—non-chain reacting." This now seems prophetic considering Grimmett's later involvement with nuclear reactors. The front of the menu might also reflect Grimmett's attitude at the time. Below Needham's name is the faux Latin phrase from the recent war; "Nil Carborundum Illegitimi"—which loosely translates to "don't let the bastards get you down," which might well have been Grimmett's own motto considering his abrupt dismissal from the Medical Research Council in England a few years earlier [68].

The M.D. Anderson records also state that he became a Counselor of the Natural Science Section of UNESCO in 1947, and his diary for that year shows him attending councilors meetings [41, 69].

While with UNESCO, he made several trips. In July of 1948, he went to Stockholm to attend a meeting of the International Consultative Committee for Radio Communications of the International Union of Radio-Science (C.C.I.R.-U.R.S.I.), an organization to which UNESCO gave annual support. At the meeting, he would have been able to renew his acquaintance with Edward Appleton, one of his lecturers at King's College. Appleton had received the Noble Prize the previous year and was active in the International Union of Radio-Science. The

5 The Unknown Years and UNESCO, Paris, 1944–1948   31

Dagens Nyheter (Daily News), one of Sweden's major newspapers, published an article on the wives of the delegates, including Mrs. Grimmett. Apparently, Grimmett had asked for the use of a grand piano because he "just cannot do without playing every day" [70].

He also traveled to Mexico City in October of 1947 with a stop in the United States. This was possibly to make arrangements for the Second Session of the General Conference of UNESCO, which met in Mexico City in November and December of that year.

During this time, he continued his association with the Hospital Physicists' Association in Great Britain, attending several of their meetings. In January 1948, he attended the annual general meeting of the HPA at the British Institute of Radiology in London and wrote several pages of notes in his diary. Of particular interest were the notes he made on Dr. Leonard Lamerton's report on the status of radiotherapy and medical physics in the United States, noting that there were very few radiotherapists in USA and that they were mainly surgeons because many of the treatments consisted of inserting radioactive needles into the patient. He also noted that there was a lack of medical physicists. Lamerton was one of the physicists at the Royal Cancer Hospital and had probably met Fletcher when he visited there in mid-1947. Grimmett also recorded a discussion, in his diary, on treatment equipment, including linear accelerators. He indicated that he thought the optimum energy for a linear accelerator would be 5 MeV [69].

Grimmett wrote his letter of resignation from UNESCO on November 9, 1948, to Dr. Pierre Auger[1]:

> I have now been away from active scientific work for four years, two years with UNESCO, and two with the Conference of Allied Ministries of Education before that, and I feel that unless I get back into scientific work soon, I shall lose the right to rank as a scientific worker.
>
> It so happens that I have been offered a most interesting post as Physicist to a new cancer Research Institute and Atomic Centre in the University of Texas, which I have provisionally accepted [4].

The group in the natural Science division at UNESCO was sorry to see Grimmett leave, and they presented him with a picture, done in pastels, depicting him in various activities in Paris, and in their concept of Texas. This picture is reproduced in part on the dust cover of this book and in color below. He is always shown wearing his flying jacket and in one scene flying a small plane while at the same time playing an upright piano, and in another scene, he is playing a grand piano on which the dates Paris 1946–1949 appear; flying and playing the piano were two of his favorite pastimes. The Eiffel Tower is crying and waving good-bye, and he is shown walking off to Texas with suitcases and scientific instruments, as well as with guns and cowboy boots with spurs. Texas is depicted as a land of cactus and

---

[1] Pierre Victor Auger (1899–1993), Director of the Department of Sciences for UNESCO 1948–1959, was one of France's leading physicists of the twentieth century. He discovered Auger electrons in 1926; he participated in the formation of the French Atomic Energy Commission (CEA) and helped organize the European Organization for Nuclear Research at CERN.

Fig. 5.1 UNESCO Good-bye pastel to Leonard Grimmett 1949 [71]

horses and pack mules and mountains, but also some skyscrapers. The skyscrapers he would find in Houston, but not the mountains (Fig. 5.1) [71].

Grimmett had planned on leaving UNESCO in early January 1949, but Auger wrote asking him to stay on a few extra days until his replacement had arrived and could consult with Grimmett. Grimmett's last day at UNESCO was, therefore, January 18, 1949 [4].

5 The Unknown Years and UNESCO, Paris, 1944–1948

It is clear that Grimmett wanted to get back into medical physics, and this is also very apparent in his letters to his wife after he arrives in Houston.

> I shall busy myself in the work, and build up a name for myself and my department, so that I can call myself once more a physicist and then hope for the future [60].

Chapter 6
Replacing Radium, 1937–1949

It is not known for sure who first had the idea of replacing the radium in teletherapy units with a more suitable and less-expensive artificial radioactive substance. Grimmett, however, had been thinking about it for some years before he went to Houston, and a case can be made that he was the first.

In the 1930s, rapid progress had been made in nuclear physics. Chadwick had discovered the neutron in 1932 at the Cavendish Laboratories in Cambridge (leading to another Noble Prize for that institution). The following year, 1933, Irene Curie and Frederic Joliot announced the discovery, in Paris, of induced or artificial radioactivity (another Noble Prize for a Curie). It was discovered that bombarding different elements with slow neutrons could induce artificial radioactivity. To obtain slow neutrons, beryllium was bombarded with alpha particles producing fast neutrons, which were then moderated or slowed down in a water bath. Curie and Joliot had used a polonium source for the alpha particles mixed with fine beryllium to produce the neutrons. In 1934, in Rome, Fermi had used radon and beryllium for the neutron source and bombarded many of the known elements. He published a catalog of numerous artificial radioactive isotopes that he had obtained [72].

This work had interested McLennan at the Radium Beam Therapy Research Unit, and since he had a large amount of radium at his disposal, a radium–beryllium neutron source was constructed. In early 1935, McLennan, Grimmett and Read published a couple of letters in Nature on the induced radioactivity that they had found in a number of elements [73, 74]. Although these proved to have no clinical interest, it does show Grimmett's interest in artificial radioactivity, and he continued to follow the literature on this subject. Among the artificial radioactive isotopes produced both by Fermi and also by Curie, Joliot and Preiswerk was sodium Na-24. In 1934, Ernest Lawrence announced the production of Na 24 by using the cyclotron and reported that from the absorption of the Na-24-emitted gamma rays in Al, Cu and Pb, a mono-energetic energy of 5.5 MeV was indicated [75]. He measured the half-life as 15 ± 0.5 h. Although the half-life was short, the high gamma ray energy interested Grimmett. Lawrence's measurement of the half-life was close to the presently accepted value, but he was off on the gamma ray energy; in fact, two cascading gamma rays are emitted with energies 1.38 and 2.76 MeV (Fig. 6.1).

In a 1937 paper in *Nature*, Eve[1] and Grimmett discussed the relative merits of radium versus high-voltage X-rays for radiotherapy purposes. They were concerned about several things including the cost of radium. The total quantity of radium in use at that time for radium beam therapy worldwide was approximately 120 g with an estimated value of £800,000 ($50 million in 2005 dollars).

The paper assessed the biological advantage to megavoltage gamma rays as compared to the X-rays produced by a high-voltage X-ray tube. In the paper, Grimmett wrote:

> Many radiologists believe that gamma-ray therapy is superior to X-ray therapy in its biological effects, and they attribute this superiority to the shorter wave-length of the gamma-rays; encouraged by this belief, they are striving after X-rays generated at higher and higher voltages, which approach the gamma-rays of radium in their nature [53].

He went on to show that even with an X-ray tube operating at 1 million volts, there would be no X-rays of that energy, whereas radium has quite a few gamma rays in the 1–2 MeV range. To match these gamma rays with comparable X-ray energies, Grimmett wrote, would require a 3-million volt tube, and he was not sure that that was possible. He was also concerned that the cost of the radium limited its use for beam therapy.

What he did realize was that the output of the X-ray tubes were much higher than could be obtained with radium, and that, therefore, treatments with X-rays could be given at extended source-to-surface distances (SSD) that gave a superior depth dose compared to that of gamma rays. Grimmett gave the following example: For a 370-kv tube at a treatment distance of 75 cm, the depth dose at 10 cm was 43 %. For a radium source at a 5 cm SSD, the depth dose at 10 cm was 11 %.

He concluded that:

> The fact is that both radium and X-ray treatments are governed by the inverse square law, and that the superior penetrating power of gamma-rays cannot be exploited unless prohibitive quantities of radium are available to make it possible to work with large radium-skin distances...

> It is possible that in a few years time the new discoveries of physics...artificial radioactivity, will find a place in radiation therapy... it is now possible to obtain gamma-rays from artificial radioactive substances with energies far in excess of anything radium

---

[1] Arthur Stewart Eve (1862–1948) was born in England and graduated in Physics and Mathematics at Cambridge. In 1903, at the age of 41, he moved to Canada as Lecturer in Mathematics and Physics at McGill University. From 1904 to 1909, he worked with Rutherford on radioactivity. He also knew Harold Wilson who was on the faculty at McGill from 1909 to 1912 when he left to help start the Rice Institute in Houston. From 1919 to 1935, Eve was Chairman of the Physics Department at McGill and Dean of the faculty of Graduate Studies (1930–1935). When he retired, he moved back to England. Rutherford had been the first Honorary Physicist at the Radium Beam Therapy Research Unit. He died suddenly in 1937, and Eve took his place. Eve also wrote the official biography of Rutherford.

emits...if it is possible to make it cheaply in bulk, it could be inserted...into a radium unit of conventional design and used for treatment in place of radium.

In the *Nature* paper, Grimmett gave radio-sodium as an example of an artificial radioactive isotope:

> Radio-sodium, for example, disintegrates with emissions of gamma rays having energies in excess of 3 million volts. This substance has already been produced in weighable amounts; if it should prove possible to make it cheaply in bulk, it could be inserted and used for treatment in place of radium. All the knowledge which has been accumulated for radium beam therapy in the past could be brought to bear on the powerful new radiation.

Although radio-sodium has a very short half-life, he suggested that the source might be exchanged daily to maintain a sufficiently high output. Little was known about the decay schemes of radioactive isotopes at that time. The energies of the emitted gamma rays were difficult to determine, and they were determined by measuring the attenuation of the rays in various metals and comparing the result to the attenuation in the same metals for a known gamma ray energy. This method, at that time, was not particularly accurate, and it was very difficult to determine whether there was more than one gamma ray and the energies of the multiple gamma rays if they were present. Radio-sodium as it was called then was radio-active Na-24, (Na^{24}) mentioned above, with a half-life of 15 h and two gamma rays of 1.38 and 2.76 MeV, respectively. This was a fortuitous radioactive isotope for Grimmett to suggest because it is very similar in many ways to the radioactive isotope that was eventually chosen, cobalt-60 (Co^{60}). Their decay schemes are very similar except for the difference in half-life (14.9 h versus 5.25 years), and they can both be produced by activation of their naturally occurring isotope, in a reactor. Not enough was known about cobalt-60 at that time for Grimmett to suggest it, but the idea was there and clearly stayed with him.

Although Eve's name appears on this paper with Grimmett, the thoughts expressed in the paper were definitely Grimmett's. Eve was the honorary physicist on the Radium Beam Therapy Research Board, Eve's name gave weight to the paper, and it was a courtesy for Grimmett to allow Eve's name to appear first. It also assured quick publication in *Nature* [53].

Even at this time, 1937, Grimmett was beginning to think about replacing radium with a more suitable and less-expensive artificial radioactive isotope, if one could be found, in the radium units of the day. It is also clear from this paper that Grimmett realized that if very large quantities could be produced, then extended treatment distances could be used, and the full advantage of the higher-energy gamma rays could be realized. Radioactive sodium was an interesting choice here and but for its short half-life would have been a serious contender. From this point on, Grimmett began to evaluate any report on artificial radioactivity to see whether a suitable candidate to replace radium had been found.

The first requirement that Grimmett would be looking for would be the energy of the gamma rays, which needed to be in the range of 1–5 MeV. The next requirement would be the half-life and how much could be made. Radium has a half-life of 1620 years so that once a treatment unit had been loaded with the source, no further radium would be needed for the life of the unit. There was, therefore, a relationship between the half-life, the amount of the radioactive isotope that could be produced and the cost involved. If the isotope had a long half-life, then there would be no need to frequently replace it, and that would enter into the cost consideration. If it had a short half-life but could be made in large quantities at a low cost then, in theory, the source could be replaced daily if necessary. This was clearly Grimmett's thinking with sodium. All he knew when he made his proposal was that sodium had been produced in "weighable amounts." Ideally, a longer half-life would be more desirable. What Grimmett was looking for was an artificial radioactive isotope with gamma ray energies of 1–5 MeV with as long a half-life as possible that could be made in large quantities at a reasonable price. In the late 1930s, no one knew the answer to the last two requirements, quantity and cost. There was one other factor that was to become very important, the specific activity of the material produced; that is the amount of radioactivity per gram of material; but in 1937, it was not of concern.

In his memoirs, Marshal Brucer, who later collaborated with Grimmett, recalled that the idea that cobalt-60 might be a suitable replacement for radium first occurred to Grimmett while he was reading Physical Review in an air-raid shelter during World War II [76]. It is known that Grimmett's house in the suburbs of London was damaged by a flying bomb and that he took refuge in his own home-built air-raid shelter during such attacks. This would have been between mid-1944 and early 1945. Brucer recalled that Grimmett told him that, at that time, he read about the 1.25 MeV radiation of a new isotope of cobalt in a short note by W.V. Mayneord. Unfortunately, this cannot be correct. Mayneord did write a paper with Cipriani on the gamma rays from cobalt-60, but this was not published until November 1947 in the Canadian Journal of Research and reported on work that must have been done in Canada before Mayneord returned to England at the end of 1946 [13]. Although there was nothing new in this paper with regard to the characteristics of cobalt-60, the authors mention in their introduction that cobalt-60 was of interest as a possible substitute for radium in certain therapeutic applications. They were not the first, however, to make this suggestion in the open literature.

During the 1930s, several investigators published reports and papers concerning induced radioactivity in cobalt. The first was by Rotblat in *Nature* in 1935, but initially, there was much confusion about the emitted radiations, especially the half-life and energies of the gamma rays; probably due to impurities in the cobalt and a competing isomeric transition with a 10-min half-life [77]. Sampson, Ridenour and Bleakney were the first, in 1936, to observe a long-lived isotope of cobalt-60 by irradiating cobalt-59 with neutrons [78]. They gave the half-life as over a year. Three other papers, however, are the most likely candidates for the ones Grimmett read in his air-raid shelter. J. Risser had published an article on "Neutron-Induced Radioactivity of Long Life in Cobalt" in October of 1937. Risser came close on the gamma ray energy. (He thought there was only one gamma ray.) He determined a value between 1.5 and 2.0 MeV (Actual values of the two gamma rays are 1.17 and 1.33 MeV.), but he did not have enough activity to measure the half-life accurately, which he approximated as 2.00 ± 0.5 years (actual value 5.26 years) [79]. A paper by Livingood and Seaborg in 1941 reported producing radioactive cobalt both in the cyclotron using deuteron bombardment as well as using a radium–beryllium neutron source to irradiate cobalt-59 [80]. They measured the energy of the high-energy gamma rays as 1.3 MeV but were off on the half-life believing a value of over 10 years was indicated. The last paper in The Physical Review that Grimmett could have read was one by Nelson et al. in 1942 that identified the half-life of cobalt-60 as 5.3 years [81].

Grimmett would have realized, however, that here was a radioactive isotope with the right-energy gamma rays and with a half-life that would allow the treatment machine to be used for several years without replacing the radioactive source. Even though the published values of 1.5–1.7 MeV, for the energies of the gamma rays were not accurately known, it was sufficiently high enough for Grimmett to realize the potential of cobalt-60. Grimmett, nor anyone else at that time, however, could have had any idea of how, how much and how much it would cost to produce cobalt-60.

Later, after the war, he would have read the paper by J.S. Mitchell in the December 1946 issue of the British Journal of Radiology [82]. This is often cited as the paper that initiated the cobalt-60 era. Mitchell specifically mentions cobalt-60 as a replacement for radium beam therapy, and he gave the half-life as 5.3 years and the gamma ray energies as 1.3 and 1.1 MeV. He also reported that it could be produced in "the pile" (nuclear reactor). Mitchell, a radiotherapist from Cambridge who was a member of the British scientific contingent that had gone to Canada during the war, acknowledged J.V. Dunsworth as co-author for these ideas, which had first been suggested in an uncirculated official wartime report in Canada in March of 1945 entitled "Application of Nuclear Physics to medicine and Biology."

But could that isotope be produced in the quantities required, and what were the quantities that were needed? No one knew. If radium was the guide, at least 10–50 curies[2] per treatment unit might be suitable. Grimmett's prewar tele-radium units were being loaded with 10 g of radium, and shortly after the war, 50-g units were in use, but no one knew whether this would translate over to cobalt-60 units. It would be sometime before a cost-effective and reliable treatment unit to replace tele-radium could be designed and built. That would not be possible for several more years, but at least a suitable radioactive isotope had been identified.

Whether cobalt-60 was on Grimmett's mind as he traveled to Houston in February 1949 is not recorded, but if it was, it was quickly driven out when he

---

[2] In 1945, radioactivity was measured in terms of curies. One curie was defined as 3.7×10^7 disintegrations per second, which was the number of disintegrations per second from 1 g of radium. Therefore, 1 g of radium could be approximately considered as 1 curie of radium although that terminology was never used. Radium was always measured in terms of its mass; all other radioactive isotopes were measured in curies. The relationship between the activity and mass of a radioactive isotope was called its specific activity in terms of activity per unit mass.

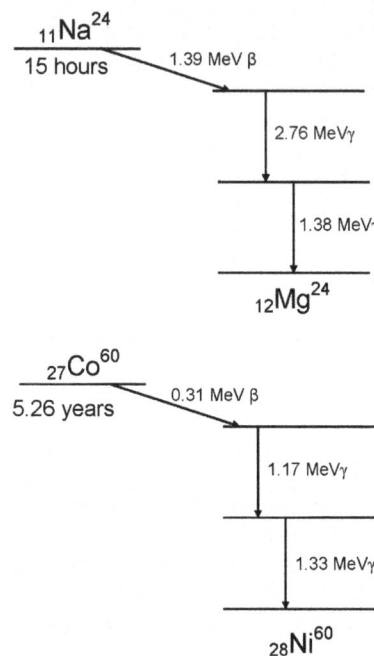

Fig. 6.1 Decay schemes for Na^{24} and Co^{60}. Except for the half-lifes, the two decay schemes are very similar

arrived at M.D. Anderson Hospital. There were more serious problems to deal with; a functional clinical medical physics group had to be established to do routine work before brand-new research could be undertaken, and the working conditions were not ideal.

Chapter 7
The Arrival, Houston, February 1949

What Dr. Grimmett expected when he arrived in Houston is not clear, but what he found did not meet his expectations, and he recorded his impressions in the letters he wrote to his wife, Norah, back in England.

Norah had stayed in England to supervise moving their possessions to America and making arrangements about their house and pets. She was to follow a few weeks later by boat. Unfortunately, there was a delay in getting her American visa, she became quite ill, the arrangements for the house created problems, she was unsure what to do with the pets, and there was a good deal of uncertainty as to the ship she would travel on.

None of this was too reassuring to Grimmett in Houston, who was struggling with his own problems, and his first few weeks in Houston were not easy.

In addition to the quote in Chap. 2 from his February 8, 1949, letter, he also told his wife in that letter:

> My first impressions of Houston and the hospital are not at all favourable (sic). The town is full of shacks and people living in caravans (trailers), and expanding at a tremendous rate. The housing problem is awful [25].

A day later he wrote her:

> But what a dump! Now that the first shock is over, I feel more indifferent, but I got an awful jar...it looks as though I shall have to make the best of it while I am here [83].

His only comfort is a photograph of his wife:

> Your picture is on the desk; it brings me comfort to look at it...

The next day, February 10th, he wrote:

> But as my earlier letters will have told you, the disillusionment I felt on arriving was terrible. First the ugly town, so different from beautiful Paris, the vulgarity of the common life, with the neon-lit cafeterias stuffed with work people in shirt sleeves... the poverty of the hospital buildings and equipment... the stories I hear of the heat and humidity of the summer to come, and above all, my sense of utter loneliness [60].

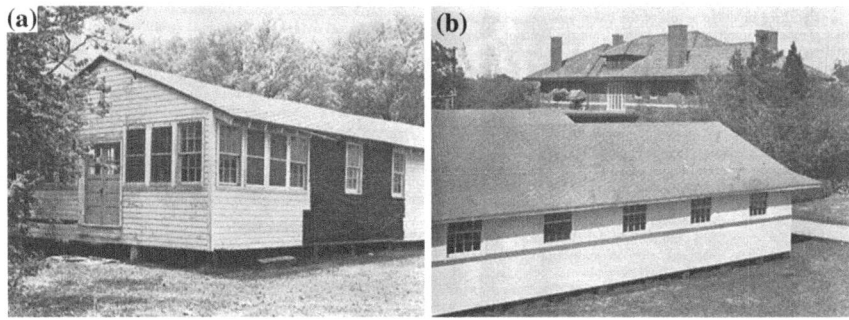

Fig. 7.1 a War surplus building from camp Wallace before renovation [84]. **b** War surplus building on the Baker Estate after renovation [84]

In 1942, when the M. D. Anderson Foundation had acquired the "The Oaks" from Rice Institute, it had to be adapted to the needs of the state cancer hospital and division of cancer research, until permanent quarters could be constructed in the Medical Center. The Oaks, located at 2310 Baldwin Street, approximately three miles northeast of the Rice campus, was a grand estate of a bygone era. The main residence was a brown brick building. The basement contained the heating plant, wine cellar, laundry room and game rooms for children and adults. On the main floor were a big reception room, a library, music room, banquet room, warming kitchen, a breakfast room, main kitchen, pastry kitchen and galleries. On the second floor were bedrooms, sleeping porches and baths. In the attic, a floor had been added to serve as storage. A wisteria-covered arbor connected the main house with the carriage house that had space for carriages, stables and automobiles. The upper floor was servants quarters. Beyond the carriage house was the temperature-controlled greenhouse that supplied plants year round for the grounds. The Rice Institute had kept the gardeners on, and the grounds were immaculately kept. As nice as all of this was, it was not suitable for a cancer research hospital, and it had to be transformed into laboratories, offices and clinics. Unfortunately, with disuse big gray wood rats had also infested the place, and a large pigeon population, with attendant fleas, lived in the stables. When the State acquired The Oaks, World War II was in progress, and it took time to convert the estate. The main residence was turned into the administrative building and also housed X-ray equipment for therapy and diagnosis. The carriage house was converted to research laboratories. A few extra buildings were constructed, and between 1948 and 1950, 13 war surplus buildings were brought from Camp Wallace, 30 miles to the south, to provide additional space [84] (Fig. 7.1).

The hospital already had several X-ray machines installed on the Baker estate and radium sources that needed to be handled and maintained by a qualified physicist. Grimmett did not know what to make of it. His wife was still in England, and he felt lonely and depressed. Until he could find a permanent residence, he was found a room in which to stay.

7 The Arrival, Houston, February 1949

> He (Fletcher) has fixed me up with quite a pleasant room in the suburbs with an engineer and his family. Some of the doctors also live here. But you know darling, how awful at my age to be a lodger in somebody else's house [25].

Grimmett sensed the fact that Fletcher could see that he was unhappy:

> I think Fletcher must have noticed it, for he has been at great pains to tell me not to be discouraged, that this building is only temporary, and that there is a great future for us here [60].

Dr. Fletcher also arranged for Dr. Jorge Awapara, a young Peruvian biochemist on the hospital staff and who was single at the time, to look after Grimmett and help him get settled in [85].

His greatest domestic concern after arriving in Houston was to find a suitable place to live and to give moral support to his wife 4,900 miles away, in making arrangements for their house in London, deciding what furnishings and pets she would or would not bring with her. There was a delay in getting her American visa, and until that could be settled, the date of her departure and the boat she would sail on were uncertain. Besides all those concerns, the weather was particularly cold in England that winter, and she became ill with the flu and other ailments.

Housing was in short supply in Houston, and it was not easy to find a suitable place. A new house was out of the question, "It was hopeless to buy a new house, as the minimum 'downer' was about $2,000 on a $10,000 house, and I simply haven't got it" [86], he wrote his wife on February 23. However, he found a house at 1742 Kipling Street situated two miles southwest of the Baker estate, but it needed painting and clean up that he undertook with the help of some of the people from the hospital. He was introduced to Texas roaches and waged a battle to get rid of them in the house. His landlady was friendly and very helpful. He went shopping for household goods but was not too impressed with what he found, and he wrote his wife to ship their Wedgwood and glass. In particular, he priced pianos and decided that there was nothing reasonably available and he asked his wife to make arrangement to have his grand piano shipped from London. There was an enclosed porch on the house, and this is where the piano was eventually placed. His neighbors at the time recalled, many years later, listening to Grimmett playing it in the evenings when he came home from work (Fig. 7.2).

He bought a used car:

> A 100 Horsepower 8-cylinder Super Ford, 1946, grey saloon…it belonged to a rich woman in Houston and has only done 18,000 miles. It fairly sizzles along, he told his wife [86].

Finally in March, his wife received her American visa, her health improved enough to travel, all their possessions that were coming with her had been packed, and she booked passage on the S.S. Charles Lykes freighter out of Liverpool bound for Houston via new Orleans and expected to arrive in Houston on Saturday or Sunday April 2 or 3.

While all of this was going on, Grimmett got some good news that his physics group at M.D. Anderson Hospital was upgraded to a separate department and

Grimmett to chairman. He was delighted and shared the news in a PPS in a letter to his wife dated March 16, 1949:

> PPS. I am 'Physics Dept' now-got my own department! Hurray! [87]

Things were beginning to look up for Dr. Grimmett. He had his own department, his wife was finally on her way to join him, he had found a house to live in, and spring had come to Houston.

His wife had told him about the cold weather in London and he replies:

> So it's cold in London? Well, you'll find it striking hot here. I'm sweating as I write this letter. You will like it, I think. The most beautiful flowers are coming out all over the town and making it look fine. Somehow you don't notice ugliness when there are flowers [88].

Grimmett had one other contact in Houston. His old friend and colleague in London, John Read, had been a graduate student at Caltech in the 1930s and had become friends with a physics postdoc there named Tom Bonner. By 1949, Bonner had become the chairman of the physics department at the Rice Institute, which was three miles southwest of the Oaks. Immediately upon arriving in Houston, Grimmett made arrangements to visit Bonner. He reported this visit to his wife in a letter he wrote on February 10.

> I met Dr. Bonner of the Rice Institute yesterday (Huxley[1] once taught there) and spent a happy evening at their house. Mrs Bonner is Czech and extremely beautiful. They want to meet you.
>
> Prof H.A. Wilson[2] (famous physicist) called into the house later (he is O.W. Richardson's[3] brother-in-law). He has retired now. It was interesting to meet him.
>
> Everyone here is kind and affable. I suppose I shall settle down in time [60].

Grimmett found a great supporter in Tom Bonner. Cooperation between the physics department at Rice Institute and the fledgling M.D. Anderson Hospital had been going on for several months before Grimmett arrived in Houston.

In his 1949 letter to Shields Warren, seeking AEC funding help, Dr. Clark mentions this cooperation. In part, he wrote:

> We now have procured the basic personnel and feel that they are most superior. Dr. Gilbert H. Fletcher, our radiologist, had five years in higher mathematics and physics before taking his medical degree...

---

[1] Sir Julian Huxley (1887–1975), Professor of Zoology, was a humanist and atheist and popularized science for the general public. In 1948, he was head of UNESCO. Dr. and Mrs. Grimmett probably met him when Grimmett worked at UNESCO but could also have met him when Grimmett was a student at King's College London since Huxley was professor of Zoology there at the same time.

[2] Harold A Wilson (1876–1965) was the chairman of the physics department at Rice and had come to Houston in 1912 when the Rice Institute was started.

[3] Owen W. Richardson was Harold Wilson's brother-in-law. They had been graduate students together at the Cavendish laboratory in Cambridge England at the turn of the century. Richardson received the Nobel Prize in Physics in 1928. He was chairman of the physics department at King's College London when Grimmett studied there.

Fig. 7.2 2006 Photograph of the house on Kipling street that Grimmett rented when he came to Houston in 1949. The porch on the right was where he had his piano and the neighbors would listen while he practiced [89]

> Our physicist, Dr. Leonard Grimmett is coming from England. He has had a number of years work with high voltage generators and was consultant to UNESCO in radiation physics...
> We have an arrangement with Rice University (sic), Houston, whereby they send a graduate student to our institution on a scholarship basis as part time to work on the application of physics to radiation therapy...
> We would like to give complete instruction in the field of medical physics as a postgraduate entity, and will be able to offer a course leading to post graduate degrees affiliated with the basic science department at Austin. We are planning to approach the Rice University (sic) Department of Physics to assist in this part of our program... [19]

It was not Rice University, but Rice Institute. Clark knew this. Perhaps, he thought calling it a university would give more weight to his request.

The arrangement with Rice's physics department to send a graduate student to the hospital had started the year before in 1948 with Dr. Tom Bonner, chairman of the physics department at Rice, being appointed as a consultant to the hospital. The first student in the program was Jasper E. Richardson.

Soon after Grimmett arrived in Houston, he outlined the work of his department for the remainder of the year. The nine-page document that he wrote was entitled "Provisional 1949 Work Plan for the Physics Section." It was submitted to the hospital administration on February 15, 1949. It began with a two-page synopsis that listed the projects to be undertaken divided into three priority groups. It included a variety of activities that would be expected for a new department:

routine clinical work, radiation safety concerns, planning for the immediate future and for the new hospital, educational activities that were to start in six months, contacts outside the hospital and a broad range of research activities. It was an ambitious program even for an established department let alone a new one just starting. Grimmett jumped in with both feet and was soon getting the program underway.

Under "Standardization of X-ray fields," he wrote:

> The best that can be done until more equipment is to hand is to build a cubical phantom of "presswood" sheets, containing slots to admit the Victoreen dosemeter...
>
> Through the kind offices of Dr. Bonner of the Rice Institute, the presswood cube is now being made in the Rice workshops.

Presswood was a commercial composite wood material used to simulate tissue because both had similar densities. The Victoreen dosemeter was an instrument for measuring radiation. Dr. Grimmett had the physics shop at Rice fabricate the presswood into a cube, called a phantom, with slots machined to hold the Victoreen dosemeter, so that he could measure the amount of radiation absorbed in a tissue-like material.

Then in a section on "Contacts with academic and industrial institutions," he says:

> It will be helpful to have the goodwill and friendly interest of the nearby institutions, such as the Rice Institute... Contact has already been established with Dr. Bonner, Professor of Physics at the Rice Institute. He has kindly offered workshop facilities and loan of apparatus until our own department is sufficiently equipped to carry on by itself [90].

It is not surprising therefore that one of Grimmett's first undertakings upon arriving in Houston was to establish a physics machine shop in the hospital. This had been a major interest of his in England. He excelled in the design of equipment, and as Read pointed out in his obituary, he required that "An instrument must not merely work: it must be well designed, elegant, and properly finished..." [3]. He designed the shop down to the finest details and planned to furnish it with surplus precision machining equipment, tools and all the necessary supplies. The list was long and quite expensive, and at Grimmett's request, Dr. Fletcher personally carried the request to Dr. Clark.

> I remember as if it were today the moment I handed the document to Dr. Clark in his office on the second floor of Captain Baker's house on Baldwin Street," wrote Fletcher in 1979. He continued, "He started perusing it and as he was flipping the pages, the look on his face became more and more puzzled. As a surgeon, he could not possibly see that all the precision tools and accessories could have any bearing on the treatment of cancer. When he had flipped the last page, he looked at me and, with a sigh, asked, 'What are we going to do with all of that?' I said that I thought the physics shop was indispensable for the development of radiotherapy. He then said, 'Gilbert, if you really think we need it, we will get it' [9].

With the shop approved, Grimmett hired E. Bailey Moore as the shop supervisor. It has been pointed out that Mr. Moore displayed the ability to return work, not only completed, but with improvements as well. A trait Grimmett surely appreciated. Bailey Moore worked with Fletcher and Grimmett on the design of the cobalt unit and the gynecologic applicators and went on to establish the shop as one of the finest such facilities in a hospital anywhere in the country. But in the beginning, it had not been easy, and after the expense of buying the equipment, money was tight. In order for Moore to immediately start work, arrangements had been made to borrow a used drafting machine and accessories. In November 1949, the company supplying the equipment wanted to be paid $45 for the equipment, which would have cost twice as much new. Grimmett tried to resolve this problem but ran into the policy that the State could not purchase used equipment. Finally, it was resolved, and Mr. Moore got his drafting machine. By January 1950, Grimmett and Moore were increasingly involved with the cobalt-60 project, and additional help in the shop was required. After discussing this with Clark, Grimmett went ahead and located a suitable machinist to join the staff but was told he had to wait on the awarding of grant money before the hiring could take place. Grimmett also asked Clark for, *"... colored help in the workshop, in the capacity of a general handy-man."* Clark told him that, *"arrangements regarding colored help if your budget will permit"* would be made [91]. This was in 1950 when the institution still had "white only" and "colored only" rest rooms and drinking fountains, etc. When all these problems had been resolved, Mr. Johnny Johnson was hired as the machinist, and he stayed with the institution in that capacity for many years; the handy-man Mr. Robert Watson was also hired. The picture below is of the physics department at this time. Grimmett is seated in the middle and behind him is Bailey Moore and next to him on his left are Johnny Johnson and Robert Watson (Fig. 7.3).

Over 50 years later, the machine shop is still in existence producing excellent precision-machined research equipment for the whole institution.

Nowhere in the first version of the research plan did he mention cobalt-60; in fact, there was little reference to radioactive isotopes or anything else that could have come under the "atomic center" that had previously been promoted. This must have crossed Grimmett's mind as he reviewed his memo because he added an additional section on "Treatment by Radioactive Isotopes," as a handwritten note.

Clinical treatment by radioactive isotopes. The Physics Department could usefully cooperate with the Clinical Department in the administration of radio phosphorus and radio iodine by assisting with the computation of uptake of radioactive substances and the estimation of dosage.

(L. G. Grimmett)
(15 Feb. 1949)

The document was produced just one week after he had arrived in Houston!

The need to get the program underway for radioiodine and radio phosphorous was pressing, and Grimmett immediately started working with Dr. C.L. Spurr of the department of medicine who was the hospital's first full time chief of clinics. Approval from the Atomic Energy Commission to use these isotopes was received by the end of March with the first shipment due in April, and Grimmett was

Fig. 7.3 Physics Department staff photograph, spring 1950. *Back row left* to *right*: Jasper Richardson (Rice Institute fellow and graduate student), Charles McLean (assistant physicist), Bailey Moore (head of machine shop), Cecil Johnson (machinist) and Robert Watson (machine shop janitor). *Front row*: Trudy Kocian (secretary), Leonard Grimmett (chairman) and Beverly Mutrux (technician) [92]

designated to play an important role in getting this program running, including the necessary training of hospital staff.

It was not as though the possibility of a cobalt treatment unit had not been discussed at M.D. Anderson Hospital, it had been, but Grimmett knew that his first task was to get a viable, reliable well-trained physics group up and running before he could do much else. The problem with cobalt-60 was that no one knew, in February 1949, what the availability and cost of it was or indeed if amounts of activity in a form suitable for a treatment machine could be produced. Newly arrived in America, Grimmett would have had little idea who to contact to find answers to those questions, although he did suggest that it would be… "Helpful to have goodwill and friendly interest of… institutions, such as… Oak Ridge Institute of Atomic Studies" [90]. He meant, of course, the Oak Ridge Institute of Nuclear Studies (ORINS), which he probably heard about from Dr. Clark who was on the Medical Review Board of ORINS at that time.

His initial outlined for the department, however, was extraordinary; in broad strokes, he laid down what he believed constituted a well-rounded hospital physics group covering the three areas of the original charge for setting up the hospital: patient care, research and education. For nearly 60 years, the basic structure of the department has not changed.

But before he could get started, however, the whole project nearly got terminated. On April 1, 1949, the Houston Post ran the following headline, "Cancer Work in Peril City May Lose Atomic Center" [21]. And this was no April fool's joke! (Fig. 7.4)

The Post wrote:

...The Texas Legislature seems about to cripple the statewide program of cancer treatment and cancer research by cutting off a building fund promised to the university's M.D. Anderson hospital for cancer research,

Besides the story, the Post devoted its editorial to the problem.

If the Texas Senate does not reinstate the $1,350,000 appropriation for construction of an atomic research building at the M.D. Anderson State Cancer hospital in Houston, which the state affairs committee has knocked out, it will lose to the state an additional $1,350,000 offered by the Anderson foundation to match state funds. Also it will lose $750,000 which has been allocated from federal sources subject to state matching [22].

The crisis pointed out how unsuitable the temporary quarters for the hospital at the Baker estate were and how the delay in constructing the new hospital in the Texas Medical Center due to the shortage of material because of the Korean War was hurting the institution. The problem was that the frame buildings of the Baker estate and the temporary wooden structures from Fort Wallace where in no way suitable for housing large amounts of radioactive materials or high-energy radiation machines. A new building needed to be constructed with massive amounts of radiation shielding without which the program could not go forward.

The committee also cut other funds, including those for the university's dental college that was to be built next to the new hospital in the medical center. An intense lobbying campaign followed to restore the funds joined by the Houston Chronicle that ran articles throughout April. Individuals who were thought to have had terminal cancer but who had been cured asked for the chance for others to be saved. The fight lasted through April with the public inundating the legislature with mail and telegrams. Finally on May 4, 1949, the senate finance committee restored the appropriation, which was then passed by a special session of the Fifty-first legislature. Clark again wrote Congressman Albert Thomas to tell him that the state had appropriated the funds and telling him he thought this placed the institution in a strong position to get federal funds especially from the Atomic Energy Commission and asking for the Congressman's advice. He concluded his letter with:

I submit this information to you to obtain your advice regarding our future strategy in achieving help to build the best in institutions in our part of the country wherein we can carry on research, education and treatment in relation to cancer patients" [20].

THE HOUSTON CHRONICLE SATURDAY, APRIL 2, 1949

Poor Place to Economize

Fig. 7.4 *Houston Chronicle* cartoon, April 2, 1949. Funds for the Atomic Building for State Cancer Research is cut $1,350,000. [93]

On March 1, 1950, Governor Alan Shivers signed the bill stating that it was, "one of the best investments the State ever made." Approximately $5,000,000 was then available for constructing the new building [23].

Chapter 8
The Cobalt Unit, 1949–1954

In 1949, the use of radioactive isotopes in medicine was new and exciting and offered great promise in the diagnosis and treatment of many diseases, not least of which was cancer. As we saw in Chap. 1, Clark saw this as a means by which the new M.D. Anderson Hospital for Cancer Research could make its mark.

Dr. Marshall Brucer was the newly appointed head of the medical division of the Oak Ridge Institute for Nuclear Studies (ORINS). At the time, Brucer was commuting between Galveston (his previous appointment had been with the University of Texas School of Medicine in Galveston) and Oak Ridge, through Houston. In May 1949, he met with Grimmett in Houston.

In Brucer's words:

> ...Grimmett was radiation physicist at Houston's cancer hospital, not yet a citizen. I had just been appointed chairman of the Oak Ridge isotope research hospital and was looking for ideas. I stopped of to see Grimmett on my way to the super-secret city of Oak Ridge and was given a complete history of all the warts on the radium bomb. Co-60 might be, Grimmett said, the answer to cancer. I invited him to Oak Ridge [76].

This was a very busy time for Grimmett, he had to setup and run the physics group at M.D.A.H., and on August 12, 1949, he revised his proposal for the future of his department. Now, he included one entitled "Proposal for the Use of Cobalt-60 in Radiotherapy."

He wrote:

> In a short paper to be given at Oak Ridge in the beginning of September 1949, details will be given of some methods of using cobalt 60 as a substitute for radium in radiotherapy of cancer [94].

He describes three areas that he would discuss: an improved cobalt-60 needle, "Cosine Law" applicators and a telecobalt unit. He outlined what he would say about the cobalt unit:

> Proposals will be put forward for methods of utilizing up to 50 curies of cobalt 60 as a mass irradiation unit. Principles of design will be discussed, with special reference to

protection of patient and operator. Designs of a machine in which the cobalt 60 can be transferred pneumatically to and from the storage safe will be shown.
Estimated costs, $25,000–$30,000 (exclusive of building).

It is clear that, at this stage, Grimmett was thinking about building a unit very similar to his radium units, simply substituting cobalt-60 for radium.

Although Brucer might have invited Grimmett to Oak Ridge, it required certain formal procedures to arrange for visitors, especially non-citizens, to go there. The University of Texas, therefore, asked for the invitation for Grimmett to visit Oak Ridge. The President of the University Dr. Painter wrote Dr. Pollard the Executive Director of the Oak Ridge Institute of Nuclear Studies on June 23, 1949, seeking permission for Grimmett to visit ORINS and take some of their courses [95]. Pollard handed the assignment over to Brucer who made all the arrangements for the visit. Since Grimmett was not a U.S. citizen, he was unable to get security clearance and so could not be hired as a consultant, but he was welcome to come and attend the courses and seminars and give a seminar of his own. Brucer was very interested in Grimmett's ideas about cobalt-60 and wrote to Clark, "We shall look forward to Dr. Grimmett's visit. I believe we can learn far more from him than we can teach him at this present time" [96]. By early August, the final arrangements had been made for Grimmett to visit Oak Ridge from August 22nd to September 2nd.

Brucer wrote Grimmett on August 12th, 1949:

> I am sending you the brochure on the Modern Physics Symposium. You have definitely been enrolled as a member of the group attending these seminars. I am tentatively arranging for you to speak to the participants of the radioisotope course late in the afternoon on Augusts 23. I believe this would be the best time to talk on the mechanism of action of ionizing radiation on living cells. On Friday, the 26th, there will be a small group of our staff, plus a few students who will be interested in a very informal seminar, and it is at this time that we would relieve you of all your knowledge of Cobalt-60 [97].

Clark had also written Dr. Lough, Chief of the Radioisotope Branch, Isotope Division requesting the use of Grimmett's signature on future orders from M.D.A.H. for radioactive isotopes, and on August 18, Lough replies to Clark saying, "It will be a pleasure to meet Dr. Grimmett on his visit to Oak Ridge and to discuss with him the various phases of radioisotope procurement and health physics" [98]. He went on to offer Grimmett any help that he could during his visit.

When Grimmett got to Oak Ridge, he and Brucer met with Paul Aebersold,[1] who was head of the isotope division of the Oak Ridge National Laboratory. Grimmett probably knew about Aebersold. Spear whom Grimmett had worked

---

[1] From 1938 to 1942 Dr. Aebersold (PhD., California, 1938) was associated with the physics and biophysics research at the Radiation Laboratory, University of California, under the direction of Dr. Ernest O Lawrence, inventor of the cyclotron. During the war he worked on various phases of the atomic energy project in Berkley, California, Oak Ridge, Tennessee, and Los Alamos, New Mexico. In 1946 he became chief of the Isotopes Branch of the Manhattan District and then chief of the Isotopes Division of the Atomic Energy Commission, which had supplied by late 1949

with in 1933 and 1935 on the radiation effects on cell mitosis had spent 1938–1939 at Berkley in California working on the effects of radiation on mammalian rat tissue with Paul Aebersold.

Brucer recalls that Grimmett initially asked for 10 Ci of cobalt-60. This was probably because it was analogous to the 10 g of radium that was then being used in the radium irradiators that Grimmett had designed before the war. Aebersold thought that a few 100 Ci might be available, and Brucer rounded it out to an even 1,000 Ci. The 10 Ci was entirely consistent with the proposal he had written the month before. Grimmett was envisioning, at that time, a short treatment distance machine, similar to the radium units for head and neck treatments, and Fletcher was very interested in such a treatment unit [76].

When Grimmett realized that 1,000 Ci was a possibility, he knew that a treatment distance longer than the radium units was possible, and he started immediately to design a unit taking full advantage of the benefits to be had from such a large amount of activity (Fig. 8.1).

With the possibility of 1,000 Ci of cobalt-60, Grimmett knew that a design similar to his pneumatically transferred radium unit would not be viable. He, therefore, turned to some of the ideas that had been incorporated into an earlier radium unit with which he had been associated, the 4-g unit at Westminster Hospital [47]. In this unit, the radioactive source (radium) was positioned in the treatment head, approximately a sphere of radius 13 cm, which was suspended, via a yoke, from the ceiling with a counter-weight mechanism (a similar sphere) to allow the treatment head to be easily raised and lowered. When the source was to be removed, the treatment head could be positioned over a safe and lowered until it docked with the safe. The source was remotely attached to a rod that was used to move it in or out of the treatment head. Grimmett envisioned a similar situation for the cobalt unit, but now, everything would be on a much larger scale. The treatment head would be a cylinder of a high-density metal approximately 45 cm. long with a diameter of 35 cm. The yoke and counter weight would have to be similarly increased in size, as indicated in the sketch. The source would be kept within the treatment head mounted on a wheel that would allow it to be rotated to an opening to obtain a beam of γ-rays for treatment or rotated $180°$ to block the rays. When the source was loaded or needed to be change, the treatment head could be positioned over a similar cylindrical safe in which the source could be transported, and the source inserted or extracted from the treatment head by a rod that could be attached to the source. Grimmett described his proposal this way [100] (Fig. 8.2):

(Footnote 1 continued)
radioactive isotopes to hundreds of research institutions, universities, and hospitals all over the world.

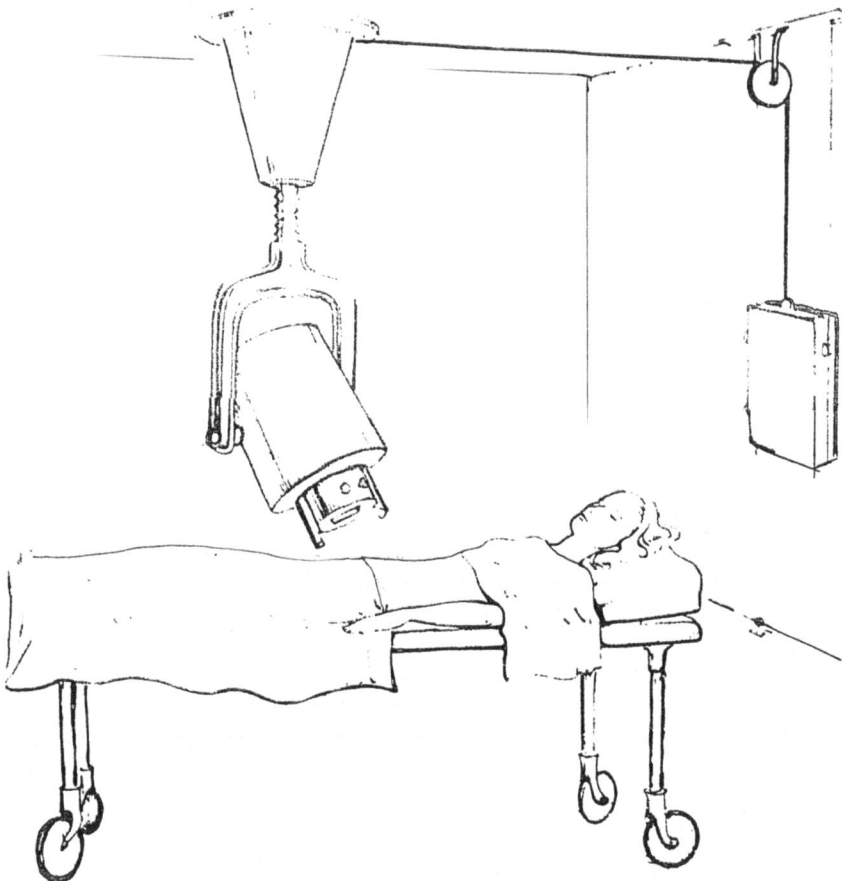

Fig. 8.1 Composite of Grimmett's 1949 sketches for the proposed cobalt unit [99]

A COBALT-60 IRRADIATOR FOR CANCER TREATMENT

The machine shown in the sketch will furnish a powerful beam of penetrating gamma radiation from a small slab of the radioactive isotope Cobalt-60, for the external irradiation of cancerous lesions.

The Cobalt-60 source will have a strength of 1000 Ci, equivalent in gamma radiation to 2,000 g of radium. It will be produced in the atomic pile, and loaded into a massive lead block, to screen off the radiation. The Cobalt-60 will be mounted on a disc of uranium, which can be rotated so as to let radiation out of the hole in the lead block when desired. In effect, this machine will be comparable to a super-voltage X-ray set working at about 2 Million V. The beam of radiation is expected however to show distinct superiority over the conventional super-voltage X-rays. The skin reaction will probably be less, and the constitutional effects on the patient smaller. It will be possible to ensure adequate safety for both patient and operator.

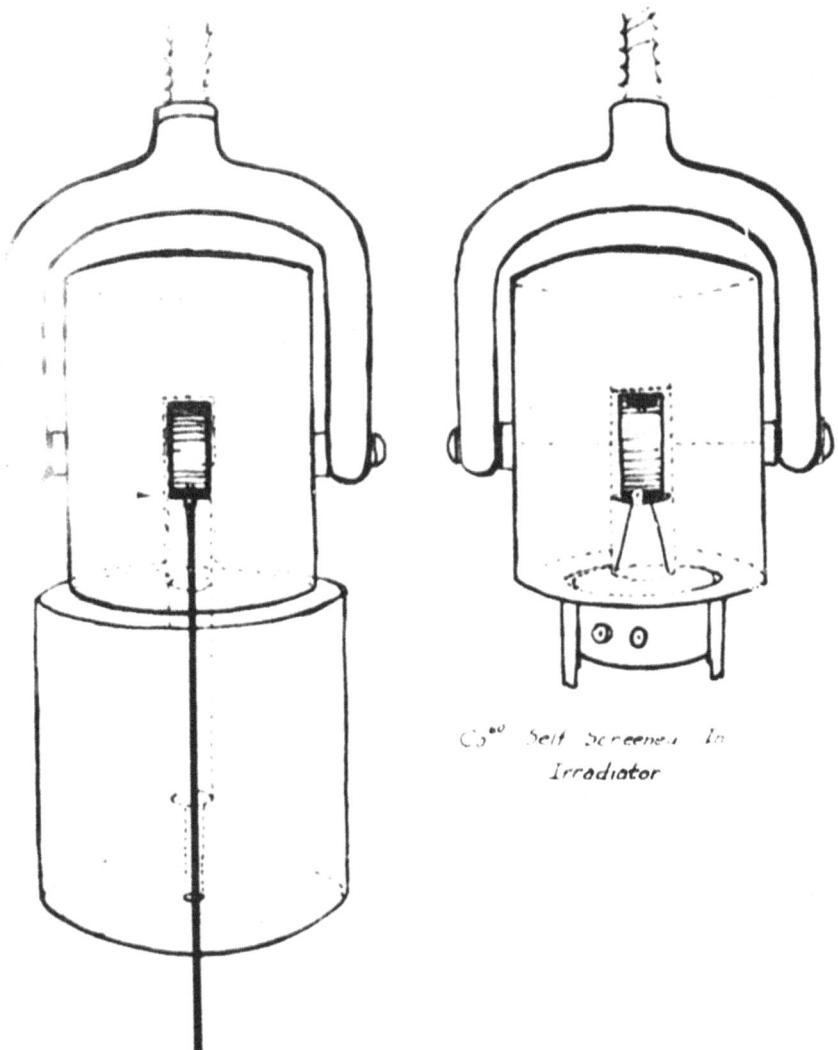

Fig. 8.2 Grimmett's proposed method for loading the unit with a 1,000 Ci source into the cobalt unit [101]

Although Brucer in his vignettes about this subject refers to a 1,250 Ci source while Grimmett refers to a 1,000 Ci source, the discrepancy is resolved in a paper published after Grimmett's death:

> The desired activity of cobalt chosen for this unit was specified to be approximately 1,250 curies. This will yield an effective curie value of approximately 1,000 curies, allowing for self-absorption of the activity within the source [102].

Sometimes, the unit, therefore, is referred to as a 1,000 Ci unit and sometimes a 1,250 Ci unit.

It was during this visit to Oak Ridge that Grimmett and Brucer started talking about a joint project between ORINS and the M.D. Anderson Hospital to produce a cobalt unit. After Grimmett's return to Houston in September, Dr. Clark wrote Brucer thanking him for, "... the excellent reception given to Doctor Grimmett in his recent visit to Oak Ridge" [103]. He went on to say, "I think the Cobalt-60 problem would be an ideal one for coordinating effort." Brucer planned to visit Houston in October of 1949, and Clark was looking forward to that visit.

Apparently, the visit went well, and cooperation between the two institutions took a further step forward.

On November 8, 1949, Brucer wrote Clark:

> I presented the general plan of cooperation between the M.D. Anderson Hospital and the Institute and explained some of the details of the problems we will encounter to the Board of Directors of the Institute yesterday. Dr. Painter (President of the University of Texas) sent a very nice letter giving the complete approval of the University of Texas, and the Board has therefore given its blessing to the proposal.
>
> I am now proceeding on the assumption that everything is cleared for us to write up a letter of agreement and to investigate how we can get the million volt irradiator bought as quickly as possible.
>
> Many thanks for your kind hospitality during my stay in Houston [104].

Ten days later, Clark wrote to Brucer saying that "the proposal for the coordinated project between the M.D. Anderson Hospital and your institution for the use of radioactive cobalt in the treatment of cancer patients," has been sent under separate cover [105].

It was now necessary for this joint proposal to be presented to the Atomic Energy Commission (AEC) in order to get approval for the production and use of the radioactive cobalt-60. A joint meeting was, therefore, setup for December 19 and 20 in Oak Ridge between M.D. Anderson Hospital, ORINS and the AEC. Arrangements were made for Doctors Clark, Fletcher and Grimmett to go to Oak Ridge and along with Dr. Brucer of ORINS to meet with Dr. Holland, Director of the Office of Research and Medicine of the AEC and Dr. Aebersold as Chief of the Isotope Division of the AEC [106, 107]. In preparation for this meeting, Grimmett sent Brucer his notes on Co^{60} along with the drawings of his proposed unit, which Brucer incorporated into booklets about the project to be presented at the meeting [108].

The proposal presented to the A.E.C. was to design, build and install a multi-curie cobalt-60 unit in the hospital of the medical division of the ORINS as a collaborative effort of both ORINS and the M.D. Anderson Hospital. The project was to proceed along the following lines:

1. A contract was to be drawn up providing for expenditure of funds supplied by M.D. Anderson Hospital for the construction of the unit. No M.D. Anderson funds were to be used either in the construction of the building to house the unit

at ORINS or for the radioactive cobalt-60. The equipment supplied by the M.D. Anderson Hospital was to be carefully identified.
2. ORINS was to construct the building in connection with the hospital of the medical division of the ORINS to house the unit.
3. The cobalt-60 was to be allocated to the M.D. Anderson Hospital free of charge for cancer research.
4. A paper transfer of the cobalt-60 to ORINS would take place for its use in Oak Ridge.
5. After the completion of the experimental studies in Oak Ridge, several possibilities were to be considered in the distribution of the equipment supplied by M.D. Anderson Hospital. These were:

   a. The equipment was to be returned to the M.D. Anderson Hospital at their expense.
   b. ORINS could take up an option to buy the equipment, the details to be spelled out in the contract.
   c. ORINS would furnish the M.D. Anderson Hospital with equipment equivalent to that originally supplied.

6. Brucer and Aebersold would arrange additional meetings with other groups to explore mutual interests in the construction of units to house multi-curie sources of cobalt-60 in order:

   a. To arrive at a proposal that would be satisfactory to the majority of interested parties.
   b. And to consult with General Electric X-ray and other interested companies regarding the possibility of constructing such a unit with the view of keeping costs low [109].

By the end of 1949, the medical division of ORINS and the University of Texas M. D. Anderson Hospital for Cancer Research had prepared the joint proposal to the AEC for the design and construction of a 1,000 Ci cobalt-60 therapy unit. It outlined in some detail the project and how the work would be divided between the two institutions [110]. They agreed to:

...cooperate in the design, construction and preliminary experimental work necessary to the production of a 1,000 curie telecobalt cancer therapy unit...
It is agreed by the two organizations that they will cooperate in testing it and measuring the physical and biological characteristics of the therapy unit.

The preliminary experimental measurements and initial therapy of cancer patients were to be done at Oak Ridge. This in part reflected the situation in Houston at the time. There were no suitable sites on the Baker Estate, where MDAH was temporary located, to put the unit. Ground breaking for the hospital's new building had not yet taken place (it would take place on December 20, 1950, one year to the day after the meeting in Oak Ridge that outlined the agreement between ORINS and M.D. Anderson Hospital). How quickly the new building could be constructed was in question because of a shortage of building material due to the Korean War. As it turned out, the space for the unit in the new building was not ready for four years.

As per the agreement reached in the December 20, 1949 meeting, MDAH was to make sufficient funds available (approximately $45,000) to cover the cost of materials and fabrication. ORINS was to make available sufficient funds for biological materials (approximately $10,000) and for housing the unit at Oak Ridge. Oak Ridge would also investigate how much shielding in the walls of the room housing the unit would be sufficient; information that was needed in designing and building the new facility in Houston to house the unit.

By the end of January 1950, Grimmett had sent a memo to R. Lee Clark, the Director of the M.D. Anderson Hospital, with the estimated cost of the materials for the fabrication of the head of the cobalt unit but not the suspension mechanism and control panel. The list included such items as steal, lead, tungsten alloy, bearings, shaft and drive mechanisms, electronics, etc. and came to $4769 [111].

It was not clear, however, where M.D. Anderson was going to get their share of the money. Houston independent oilman and owner of the Shamrock Hotel in Houston, Glen McCarthy, came forward with a suggestion. He was sponsoring the Shamrock Charity Bowl Football Game and Dinner in December 1949 in part to benefit the Damon Runyon Memorial Fund for Cancer Research. Since the Damon Runyon fund had a policy of making grants to institutions in the city where the funds were raised, McCarthy suggested that Dr. Clark apply to the fund for help in developing and building the telecobalt unit. The football game was played in Rice Stadium on December 17, 1949, under very rainy conditions, which somewhat limited attendance, but on March 16, 1950, Mr. McCarthy standing in for Walter Winchell, founder and treasurer of the fund, presented Dr. Clark a check for $16,000 [112]. Clark noted that this gave some financial support and a much needed spark of enthusiasm to the project.

Other support was also sought. On February 4, 1950, Grimmett prepared information about the unit, and a diagram (probably similar to the sketch shown above) to be sent by Dr. Roy Heflebower, assistant director and administrator of the hospital, with a cover letter to the Cancer Control Fund [100].

As suggested by the group in the December 20, 1949, meeting, ORINS and the Isotopes Division of the A.E.C. called a meeting on February 15, 1950, in Washington D. C. specifically to discuss and solicit designs for a cobalt-60 irradiator. Gilbert Fletcher and Leonard Grimmett attended the meeting on behalf of M.D. Anderson Hospital for Cancer Research along with thirty-one other attendees from around the country and Canada [113]. About half were radiologists, and one-third physicists, and the rest from various government agencies and industry [114].

Dr. Paul Aebersold and Dr. Allen Lough opened the meeting with a discussion about the limited supply of cobalt-60 and what was going to be available from the Oak Ridge reactor. It was proposed that if a common design of a cobalt-60 unit could be agreed upon, it might be possible to reduce the cost of producing them. Lough presented the list of cobalt-60 sources being prepared at Oak Ridge including their expected activity, specific activity and physical dimensions. He thought that the ultimate maximum specific activity expected from the Oak Ridge reactor would be 2 Ci per g. The cost of a 500 Ci source would be $2,600.00. Grimmett's design called for a 1 × 4 × 4 cm cobalt- 60 source of approximately 1,000 Ci. This was critical because it meant that the specific activity of the source had to be around 7–8 Ci per g, far in excess of the 2 Ci per g promised by Dr. Lough. Both Lough and Aebersold stressed the importance of enclosing the cobalt metal in an aluminum container because the cobalt metal oxidizes within the reactor, forming a highly radioactive layer of cobalt rust in the form of a fine white powder. It was apparent, therefore, right at the beginning of the meeting that the supply and availability of suitable cobalt sources from Oak Ridge for treatment machines was in question. In fact, Aebersold went further and said that future deliveries of cobalt-60 could in no way be guaranteed because irradiations for medical purposes did not have high enough priority to be ahead of other projects. The clear implication was that the top priority for the reactor was military use.

Dr. M. H. Thomas, who was chief of the Radioisotope Branch at the Canadian A.E.C. Chalk River reactor in Ontario Canada, then described the situation in Canada for the supply of cobalt-60 that was in stark contrast to that at Oak Ridge. At Chalk River, specific activities in the range of 2.3–6 Ci per g were available, and sources that were currently being irradiated would have higher specific activities, some in the range of 27–40 Ci per g and others in the range of 18–33 Ci per g, yielding a total activity of 3,500 Ci. There were two cylinders of cobalt, 3.8 cm in diameter and 3.8 cm high, being activated for six months that were expected to have a specific activity of 14 Ci per g yielding a total activity of 2,520 Ci that he estimated as being equivalent to 1,200 Ci of unshielded cobalt-60. The cost for these sources would be in the range of $3000–$4000 each. He reported that the Canadian AEC would consider the activation of any specific cobalt source submitted to them provided that they met the Canadian requirements for insertion into their reactor, but that deliveries of cobalt-60 could not be guaranteed as civilian production took second place, again presumably to military use.

Dr. Thomas also announced that the Eldorado Company in Canada was planning to market a 1,200 Ci cobalt-60 unit for medical purposes at a cost of $25,000 and showed a preliminary drawing of the unit, but no further details were given.

Aebersold reminded the meeting that under the existing U.S. laws at that time, all purchases of radioactive isotopes from outside the U.S.A. had to be approved by the U.S. A.E.C.

The rest of the morning session was a series of presentations concerning different designs for possible cobalt units. Dr. Marshall Brucer as Chairman of the Medical Division of ORINS reported on the collaboration that had already taken place between ORINS and MDAH. He outlined the principle requirements that would have to be met in any design of a large cobalt-60 irradiator. They were:

1. Simplicity and safety in loading the cobalt-60 into the machine.
2. Adequate protection for operators.
3. Simplicity of arrangement for turning the γ-ray beam "on" and "off".

4. Flexibility of machine, that is, ability to raise and lower the head and to orient the beam in any desired direction (a movement of 120° in one plane was sufficient.).
5. Provision of diaphragms for changing the size of the treatment field.
6. Provision for fitting the usual clinical accessories for radiotherapy, such as light-localizer, back-pointer, range-finder, etc.

He used Grimmett's design to illustrate these points. Next, Dr. Grimmett talked about the penumbra produced using a 4 cm^2 source at a 50 cm treatment distance, which was the dimension that he had used in his design. He stressed the desirability of obtaining as small a source size as possible. Later that evening, several of the participants at the meeting met in Dr. Brucer's hotel room to continue the discussion on source size. A source diameter that was equal to that of a half-dollar (3 cm) was considered too large. The diameter of a dime (1.8 cm) was considered too small. According to Brucer, it was at that moment that Harold Johns (from Canada), who had just paid a bell boy for some libations that had been brought to the room, took a quarter out of his pocket and suggested its diameter would be ideal; it happened to be a Canadian quarter (diameter 2.3 cm). This was smaller than Grimmett's original suggestions of a 4 cm^2 source, which meant that if the activity was to remain the same, a higher specific activity would be required. It made the possibility that Oak Ridge could supply the source even more remote.

The morning session concluded with five other designs being shown. There was one plan for a rotational therapy machine for which there was expressed a lot of support, but the plans that suggested fixed horizontal beams received little interest. Harold Johns discussed his design for the unit for Saskatchewan University in Saskatoon, Canada, noting that it was very similar to Dr. Grimmett's design.

It was, therefore, apparent that Canada had two machines under construction: the Eldorado Company machine and Dr. John's machine in Saskatoon.

Some interest was expressed in a lower activity machine that would be used at a shorter treatment distance, making it similar to the existing radium units. Such units would require only 10–50 Ci that could be more easily supplied by the Oak Ridge reactor. (This was the kind of machine that Grimmett had originally suggested.) These machines would be used primarily for the treatment of head and neck cancers, and not surprisingly, Dr. Fletcher expressed an interest in a machine of this type. He had been very impressed with the treatment of head and neck cancers with a 10 g radium unit at the Royal Cancer Hospital when he visited there in 1947. He had become good friends with Dr. Lederman who headed up that program, and Fletcher had decided to make the treatment of head and neck cancers a prime goal of his program at MDAH. However, there was little support for such a unit among the other participants.

The afternoon session gave the representatives of industry the opportunity to express their thoughts and ideas.

Dale Trout from the General Electric X-Ray Corporation expressed his company's interest in building cobalt-60 units. He compared the cost of mega-voltage X-ray units, $68,000 for 1 million volts and $120,000 for 2 million volts. This could be compared to the estimated cost of a cobalt-60 unit that had been cited, in the morning session, of $25,000 without the radioactive cobalt.

The representative for Kelley-Koett Manufacturing Company offered a suggestion for the design of a cobalt-60 unit, as did the represented from Tracerlab, Inc.

Dr. Brucer then asked for a summary of the basic principles for the design of a 1,000 Ci unit. It was agreed after some discussion that with the cobalt unit in the "off" position, the radiation intensity on the surface of the unit should not exceed 0.3 r per week.

The output expressed as the dose rate on the skin (the dose build-up at 0.5 cm depth was not yet fully appreciated) should be 50–100 r/minute, although 20–50 r/minute would be adequate.

Transmission through the diaphragm should be no more than 3 % of the full beam, and the general radiation field over the body of the patient outside of the direct beam should not exceed 0.1 % of the direct beam.

The interest in rotational therapy was a surprise, especially to Dr. Grimmett. The provision for a rotational unit was not an essential feature of his design but would be an interesting variant.

When Dr. Fletcher and Dr. Grimmett returned to Houston, they reported on the meeting noting that the group at the meeting had narrowed the choice of a design to two units—M.D. Anderson Hospital and Tracerlab's [115].

Grimmett then summarized the effect that the meeting would have on the design of his unit. First, he addressed the question raised by the problems associated with the supply of cobalt-60. He suggested that the Canadian AEC should be asked, through Dr Brucer, whether they would activate eight pieces of cobalt, each $2 \times 2 \times 0.25$ cm [116].

Originally, Grimmett had not been sure that the expense of using tungsten as the main shielding material as compared with lead could be justified. He now believed that with the meeting fully in support, tungsten should be used. Some consideration was given to using uranium as the shielding material, or at least a combination of tungsten and uranium. Clark recalled that: "For a while we had wanted to try uranium… and that brought the FBI down to investigate us." The uranium idea was quickly dropped, and it was decided to make the unit out of the tungsten alloy, Hevimet. This was manufactured in England and was the same material that Grimmett had used for shielding in his radium units. Again Clark recalls:

…we (the United States) were about to go into the Korean War and the government was keeping all the tungsten because they made rotor blades for the jets out of it. So we formed the Medical Importation Company. James Anderson helped us get it done through the Anderson Clayton Cotton Company… he was Mr. M. D. Anderson's nephew and really loved our hospital [117].

Although Grimmett had originally planned for his unit to have 1 % transmission through the field-defining diaphragm, he decreased the thickness to give 3 % transmission as recommended by the meeting. This resulted in a decrease in the weight of the diaphragms, which would be an advantage since they would have to be inserted and removed by hand. It would also require less material, decreasing the cost of the unit.

He reported that Dale Trout of G.E. had indicated that G.E. would consider fabricating a complete unit, under their name, but Grimmett was concerned about losing control of the project. Perhaps his experience with the Radium Beam Therapy Research group had soured him on cooperative efforts. He wanted G.E. to respect the hospital's claims to the original design and for the hospital to have the publication rights; if they agreed to these requirements, there might be an advantage to having G.E. build the machine. His fears would prove all too true, however.

Dr. Grimmett presented his design in a paper to MDAH.'s annual symposium in May of 1950, entitled "The Use of Cobalt-60 in Medicine" and a wooden mock-up of his 1,000 Ci irradiator was displayed at the meeting [118, 119]. This paper was then published in the 1950 winter (Oct–Dec.) issue of the Texas Reports on Biology and Medicine as, "1,000 Ci Cobalt-60 Irradiator" [120]. This is the first published paper on the design of a cobalt-60 unit.

The paper was a reprieve of his 1937 paper. He again showed the comparison of the radiation spectra of a high-voltage X-ray tube and radium, but now, the gamma-rays of cobalt-60 were included in the figure. By this time, Grimmett knew all the appropriate characteristics of cobalt-60. The values he quoted for the gamma-ray energies, half-life and exposure rate constant were very close to the accepted values in use today. The unit, however, was a far cry from the simple suggestion he made in his 1937 paper that the artificial radioactivity, "... could be inserted...into a radium unit of conventional design and used for treatment in place of radium." With 1,000 Ci, the source could be moved further away from the patient surface than 5–10 cm of the radium units; Grimmett chose 50 cm. This unit would, therefore, have a superior depth dose than the kilo voltage X-ray machines then in use, fulfilling one of the advantages he had suggested in the 1937 paper. In 1950, he wrote, "...Cobalt-60 may be considered 'equivalent' to a 2 MeV X-ray tube." He also designed the unit with a small source size, a 2-cm cube, arguing that with the extended treatment distance and a smaller source size, the radiation beam produced by the unit would have a much smaller and well-defined penumbra; something the radium units did not have. He also understood that it would be inherently dangerous to move 1,000 Ci from storage safe to the treatment unit pneumatically, as he had done with his radium units.

> The pneumatic system of propelling the radioactive material by air pressure to and from a storage safe was considered and rejected because it may on rare occasions break down. A breakdown with 1000 curies of Cobalt-60 would be intolerable.

The new unit, therefore, was self contained with sufficient shielding to make it safe to work around while setting up patients with the leakage radiation not exceeding the permissible dose rate of 0.3 r per week.

In the paper, he alludes to the problem of activating a small volume of cobalt to the high activity levels required and concluded that: "To get the desired activity into this small volume, it will be necessary to use a high neutron flux, such as is available in the Canadian pile at Chalk River." This turned out to be a serious problem and delayed by several years the beginning of clinical use of Grimmett's machine.

Although a 1,000 Ci had been suggested as the activity of the source at the initial meeting between Grimmett, Brucer and Aebersold the previous year in Oak Ridge, the fact of the matter was that 1000 Ci was not available and certainly not in the small size required for the cobalt unit nor could Oak Ridge produce such a source in a reasonable amount of time. And this problem had been known almost from the beginning.

His concluding paragraph is of interest.

> It is our eventual hope to produce a simple, cheap, and reliable machine, needing no servicing or replacement, apart from the replenishing of the source every five years or so, which will enable monochromatic gamma-rays to be tried for the first time in cancer treatment. The cost is difficult to estimate at this stage, but will probably be in the region of $30,000. It would seem to be a sound way of using atomic products, which should bring the benefits of high-voltage radiation within the reach of the ordinary hospital.

Although overly optimistic, especially with regard to the cost, Grimmett's predictions about the use of cobalt-60 units proved to be true. Thousands of units have been built and used worldwide, and millions of patients have been treated on them. Hundreds are still in use worldwide.

In July 1950, Fletcher attended the Fifth International Cancer Congress in France and presented a paper "A 1,000 Ci Cobalt-60 Irradiator" coauthored by Grimmett and himself [121]. This paper was published in the proceedings of the conference in 1953 and was basically a rewrite of Grimmett's paper. Fletcher then went on to the Sixth International Congress of Radiology in London and again presented the paper on the cobalt irradiator. Grimmett had wanted to attend this meeting, and in August 1949, he had written his friend Binks at the National Physical Laboratories in England that he hoped to be there, but he was far too busy to make the trip [122].

Grimmett continued on the design of the unit. In June of 1950, he attended a radiobiology meeting (another of his interests) at Oberlin College with Brucer and then went on to Cincinnati to look over some teletherapy hardware.

Brucer was also working on getting the cobalt source irradiated in the Canadian reactor at this time. He and Grimmett had finalized the source configuration as $2 \times 2 \times 1$ cm made up of four individual wafers of cobalt each $2 \times 2 \times 0.25$ cm. This was half the size of Grimmett's original suggestion of a $2 \times 2 \times 2$ cm source. Such a large source, however, would have too much self-absorption, so the source height was cut in half, which again required the specific activity to be increased. Obtaining high-grade Co^{59} was hard enough, but nothing compared to the bureaucratic nightmare of the AEC and their concerns for secrecy and rigid import–export regulations before the stable Co^{59} wafers could be delivered to the Eldorado Mining and Refining Company Limited, the crown company that operated the Chalk River reactor for the Canadian government. In June 1950, three sources were loaded into the Chalk River reactor: one for Harold Johns in

Saskatoon, one for Cipriani of the Canadian National Research Council, and the ORINS/MDAH source. The anticipated time for this source to reach 1,250 Ci was 10 months.

By July 1950, the Oak Ridge Institute of Nuclear Studies and the University of Texas had formalized the contract for the fabrication testing and use of the cobalt-60 irradiator, and a contract had been let to General Electric to build Grimmett's machine for $27,000 [123]. (It was learned later that G.E. put a further $40,000 into the project.)

The Korean War had started a month earlier in June 1950, which immediately increased the building costs for the new hospital, and when bids were received in September 1950, the lowest bid was approximately $2,500,000 higher than the $5,000,000 available. The plans had to be modified, and a contract was not awarded until October 1950 for $5,242,104 with the contracting firm agreeing to build the items deleted from the original plans at the price quoted in the original bid as additional funds were made available [124]. Construction officially began December 20, 1950, when groundbreaking ceremonies were held at the institution's future site in the Texas Medical Center [125]. Dr. Grimmett attended the ceremonies with great interest. The first part of the construction would be the basement with the heavily shielded walls, which he had designed, for the proposed cobalt-60 unit as well as other treatment machines. In order to reduce the thickness of the walls "heavy concrete," ordinary concrete to which a high-density aggregate has been added, was used. For the aggregate, Grimmett chose ilmenite imported from Canada. Until this building was completed, the cobalt unit could not be moved to Houston (Fig. 8.3).

Much of the unit was to be fabricated out of a tungsten alloy known as Hevimet, the same material he had used in England to shield the radium units. On March 27, 1951, G.E. wrote to Grimmett that they were having trouble machining the Hevimet and that some design changes might be necessary and that they were looking forward to his visit to Milwaukee the next week when they would give a progress report on the unit [127].

Grimmett made that visit on April 4 and 5, 1951, with H. Kerman the radiation oncologist from the University of Louisville Medical School who was on loan to ORINS. Kerman had taken the basic radioisotope course at ORINS in early 1949 and had got to know Brucer when he visited Louisville later that year. As the cooperation with MDAH progressed, Brucer realized that he would need help at ORINS with their part of the contract and asked Kerman to join him. Kerman was able to get one year's leave of absent from the medical school and joined Brucer at ORINS in the spring of 1950. The one year stretched into two. Grimmett reported in detail on this visit.

He said:

> We found much to praise, and little to criticize in the progress which G.E. has made. Our objections were carefully weighed, and modifications proposed to meet them. [128]

He was impressed by the small size of the unit that had resulted in the use of Hevimet. But he and Kerman did not like the positioning of the mechanism for

8 The Cobalt Unit, 1949–1954

Fig. 8.3 Groundbreaking for the new hospital, December 29, 1950; Grimmett, wearing glasses, can be seen immediately behind the lady on the front row looking to her right. The feather of her hat appears alongside Grimmett's face [126]

rotating the source that was planned for the side of the unit. They believed that this would interfere with the clinical setup for some patients. Grimmett, therefore, proposed a one-sided suspension mount with the source-rotating mechanism placed on top of the unit. Grimmett drew a sketch of the situation and his solution of the problem for his report on the visit. This was eventually the solution that was used. (see fig. 8.5 on page 71.) He reported at length on the mechanism that rotated the source wheel, since this was the way the unit would be turned on and off and had to be, as far as possible, fail-safe. GE had built a full-sized model of the rotating mechanism, and it had been under automatic test, day and night, for several months without failure when Grimmett and Kerman made their visit. Grimmett fully understood the problem of the safety issue in using radioactive material in a teletherapy treatment unit. Since the radioactive decay is continuous, there is no way of turning the emitted radiations on or off. When not in use, the radioactive material must be so situated that essentially, no radiation escapes the unit, but when in use, the radioactive material must be so situated that the emitted radiation can pass through a well-defined aperture. Grimmett's design for this machine had the cobalt-60 source placed on the circumference of a disk of Hevimet that when rotated would align the source with the aperture for the "on" position or rotated 180°, where the source would be completely surrounded by shielding material for the "off" position. This was to be done by an electric motor. But what would happen if there was a power failure with the source in the "on" position? Grimmett described the mechanism to handle this situation:

Fig. 8.4 Photograph of the control cabinet [129]

Based on previously agreed ideas, it consists of a small geared-down electric motor, coupled through a magnetic clutch to the Hevimet disc carrying the cobalt source. In the "on" position the motor continues to run, the clutch slipping continuously, thus holding the Hevimet disc positively against a fixed stop. When the motor is switched off, the magnetic clutch is de-energized and a coil-spring in the housing brings the Hevimet disc back to the "off" position.

He made a few more suggestions to eliminate the source wheel from hitting the stop to strongly, which he thought might over many uses loosen the source. But overall:

...We were much impressed with this mechanism. It was positive and definite in action, and safe in case of power failures.

They also liked the control panel:

This was extremely simple and satisfactory, consisting of a small steel box with buttons for starting and stopping the motor driving the Hevimet disc, and indicator lamps [128].

Later, an electric timer replaced the stop button, and an interlock key was added. With the key turned on and the desired treatment time set, the timer could be turned on, but nothing would happen, and the green safety lamp at the top left would remain on until the "irradiate" button to the left of the timer was pushed (Fig. 8.4).

Immediately, the green light would go out, and the red "irradiate" light would come on, and the source disk would rotate into the irradiate position, at which point the timer would start. The timer would count down to zero and automatically turn the unit off by disconnecting power to the source disk motor, and the return spring would rotate the disk to the off position. At which point the lamps would reverse, the "irradiate lamp would go off and the green safety lamp would come on." The whole system was interlocked for safety; turning the key off during treatment, turning the timer off during treatment, opening the door into the treatment room would all cut power to the motor, and the mechanical spring would return the source to the off and safe position. It was elegant and simple and reliable.

The project was moving along, and the unit was to be ready for delivery by June 9, 1951. However, the cobalt-60 sources were not yet up to full activity, and a controversy over the sources was brewing. There were frequent rumors that the Eldorado Mining Company was offering cobalt-60 sources and teletherapy units for delivery in 1951. But only three high-activity sources were known to be in preparation. One each for the two Canadian treatment units (Harold Johns and the Eldorado Company's) and the third for the ORINS/MDAH machine, and the ORINS/MDAH group became very concerned; so much so that Kerman and Brucer went to Ottawa to sort out the situation where they were assured that they were going to receive their source as planned. They took the opportunity while in Canada to go to Saskatoon and visit with Harold Johns and Sandy Watson, the radiation oncologist, to see John's cobalt-60 unit.

Kerman reported that:

> The unit's mechanism was very similar to that designed by Grimmett. The head was larger since Johns was using lead shielding and Grimmett had specified Hevimet. Johns collimating device seemed superior to the heavy cones that were... designed for the Grimmett unit [130].

The ORINS/MDAH sources were calculated to have a combined strength of only 800 Ci in April 1951 and would need additional irradiation but when they were removed in June their activity was found to be only 650 Ci, and it would take an additional 150 days to reach the desired strength of 1,250 Ci. It was, therefore, decided to leave then in the reactor for another 6 months to come up to a higher activity. Grimmett had received letters from Mitchell (the radiation oncologist and author of the 1946 paper in the British Journal of Radiology that first suggested cobalt-60 as a replacement for radium), and Freundlich (a medical physicist) concerning their iridium-192 teletherapy unit in Cambridge, England,[2] and their problems with low activity sources. Grimmett knew these problems from his days

---

[2] The iridium-192 teletherapy unit at the Addenbrooke's Hospital in Cambridge, England was the first teletherapy unit to use artificial radioactivity and was the first such unit to clinically treat patients [131, 132]. Freundlich was the medical physicist and J.S. Mitchell was the Director of the Department of Radio therapeutics. Iridium was used because, at the time, large sources of cobalt-60 were not available in England. It did not have as high gamma-ray energies as cobalt-60 (400 keV versus 1.25 MeV) nor as long a half-life (74 days versus 5.26 years) but high specific activities could be obtained after only a few months irradiation in the reactor. So there were two identical sources, one being in use in the unit treating patients, while the other was in the reactor and the sources were exchanged every four weeks. This unit had an output of 16r/min. at a source to skin distance of 8 cm and in performance was similar to an 8 gm radium teletherapy unit. In effect this was a unit very similar to the pre-war radium teletherapy units but with the radium replaced with radioactive iridium-192, as suggested by Grimmett in 1937.The first patient treated by this iridium unit was in May 1950, over a year before the first cobalt-60 patient.

working with low-output teletherapy radium units and wanted as high an activity for the cobalt unit as possible.

The two Canadian sources were also removed in June 1951, and both found to have an activity of approximately 800 Ci. This allowed John's unit to be operating clinically in Saskatoon by November 1951. The other source was incorporated into the Eldorado that was delivered to London Ontario and went into clinical service even earlier on October 1951.

In early 1951, the need to hire a physicist to help with the measurements on the new machine at Oak Ridge became apparent, and Grimmett turned to Jasper Richardson who had been the first graduate fellow from Rice to the physics department at M. D. Anderson Hospital.

Richardson had graduated from Rice with a Ph.D. in physics in 1950 and had taken a position as assistant Professor in the physics department at Alabama Polytechnic Institute (now Auburn University) in Auburn, Alabama. He may not have been entirely happy there. In January 1951, Grimmett wrote him a letter sending him a copy of a paper on the use of scintillation counters in radiotherapy, which Grimmett had presented at the Southwestern Section of the American Association for Cancer Research, held in Austin the previous December, since it contained work jointly done with Richardson, while he was a graduate fellow at MDAH. Grimmett had also put Richardson's name on the paper. In the letter, Grimmett explained that he had hoped to use such scintillation counters to investigate leakage radiation along the joints between metal blocks but that the scintillation counters turned out not to be suitable and Geiger counters had been used instead. This was an important study for the design of the cobalt unit. It proved to be impossible to cast a single piece of tungsten alloy in the dimensions required, so five pieces were used and machined to ordinary workshop tolerances and bolted together, and the experiments had shown that the leakage between such machined surfaces was negligible. The study did show, however, that there was considerable scattered radiation at the end of the treatment cone with the source in the "off" position. The wheel holding the source was a rotating disk with straight sides paced in a cavity in the shield that also had straight sides. The source wheel was, therefore, redesigned with a step on each side, and the shield modified to accept the step. This proved to reduce the scattered radiation at the end of the cone to tolerable levels. Grimmett had undertaken this work, using a 2 Ci cobalt-60 source that the hospital had acquired, with Robert J. Shalek,[3] the graduate fellow from Rice who had replaced Jasper Richardson. Grimmett ended his letter to Richardson with:

---

[3] Doctor Robert Shalek was the second Rice Physics Fellow in the department of physics at MDAH from 1950 to 1953. He spent a year at the Royal Cancer Hospital in London training under W.V. Mayneord returning to MDAH in 1954. He became chairman of the physics department in 1961 a position he held until 1984.

The matter of your returning to work with us has been brought up and has met with a favorable reception. I shall follow it up and see if we cannot get some definite proposal to put to you before very long [133].

Subsequently, Richardson was asked whether he would be interested in taking a fellowship at the Oak Ridge Institute of Nuclear Studies in the summer to work on the new cobalt unit. Richardson was interested, and Grimmett wrote him on March 19, 1951, to say that he was, "...glad to hear you would consider... the fellowship" [134]. He promised to send Richardson some references on cobalt-60 and also noted there might be a delay in the project and that it could be August before they started.

In the middle of these discussions with Richardson, Grimmett made the trip to Milwaukee with Dr. Herbert Kerman to review the progress being made on the construction of the unit so that it was not until April 26 that Grimmett sent Richardson the list of references that he believed would be helpful regarding the cobalt-60 program (see appendix C). He also outlined the problems that needed to be solved first.

These were:

1. Loading the cobalt source into the unit. He and Shalek would work on this in the next few months, again using the 2 Ci source.
2. Measurements of the beam in a water tank. He and Moore (the engineer at MDAH) were working on a remote control unit for moving the ion chamber in water.
3. Measurement of electronic buildup
4. Estimates of "effective" wavelengths at depth in the water phantom.

He added a P.S that he was planning to spend August at Oak Ridge [135].

The delay in the ORINS/MDAH source required some revisions in the plan. The machine was shipped from Milwaukee to Oak Ridge, and since Herbert Kerman and Jasper Richardson were ready to test it, an arrangement was made to borrow a 200 Ci source that had been prepared at Oak Ridge for Dr. Max Cutler of the Chicago Tumor Institute.

Just as things seemed to be going smoothly and the completion of the project was in sight, a disturbing incident occurred. The May 28, 1951, edition of Newsweek was published. The major story in its Medicine section was entitled "Cobalt-60 Therapy." In the article, it stated that:

...the medical division of the Oak Ridge Institute of Nuclear Studies and the General Electric X-Ray Corp. of Milwaukee are now cooperating in designing and testing a 1000-curie radiocobalt therapy unit, which has been authorized by the Atomic Energy Commission.

Only at the end of the article did it say:

When its safety has been determined, the unit will be installed at the M.D. Anderson Hospital for Cancer Research, Houston, Texas. There a series of long-range studies will be made, pointing to the development of special techniques for irradiating deep-seated tumors [136].

Nowhere in the article was anyone from MDAH quoted or given credit for the idea for the unit, not Grimmett, nor Fletcher nor Clark. It read as though the concept had been developed by ORINS and G.E. with quotes from Marshall Brucer of ORINS and Dale Trout for G.E. It seemed that Grimmett's worst fears had been realized and that he and the hospital had lost control of the project and that ORINS and G. E. had taken credit for the design of the unit and had published, in a leading national news magazine, the existence of the unit before the hospital had had a chance to do so: the two points that Grimmett had insisted upon before entering into a contract with G.E. to build the unit.

Clark had another major concern; no mention was made in the article about the financial support of the Damon Runyon fund to the project. This would have violated the agreement of the hospital with the fund that any publicity about the unit would acknowledge the support of the fund. Clark knew that Walter Winchell would read the account and might take it as an affront that the fund had not been mentioned thereby jeopardizing any further funding. He immediately wrote Winchell explaining that the hospital had no knowledge about the article and that they were as surprised by it as anyone. He sent a copy of the letter to Dale Trout.

But by the time, Trout wrote Clark in response to this letter Dr. Grimmett was dead!

It is ironic that on the same date as the publication of the Newsweek article, Monday May 28, 1951, the Houston Chronicle announced Grimmett's death. The headlines were:

Doctor Grimmett, Cancer Expert, Dies Suddenly
  Dr. Leonard G. Grimmett, 49, eminent physicist whose work in cancer research at M.D. Anderson Hospital, opened a whole new field of treatment of cancer, died of a heart attack at 1:10 a.m. Sunday at his home, 3238 Ewing [137].

A full obituary followed.

What caused the heart attack is unknown, although there was a family history of heart problems. For the two and a quarter years, he had been in Houston, he maintained a hectic work pace, and his relationship with Dr. Fletcher who had brought him to Houston had badly deteriorated. At the time of his death, his wife was on a ship going back to England to look after her sick mother. Grimmett never liked to be alone, and he had made arrangements for a married couple on the hospital staff, Dr. Stella Booth and her husband who were close personal friends, to stay with him while she was gone. Had he read the Newsweek article and had that contributed to his death? Although unlikely, it is not impossible. Copies of Newsweek are sent to libraries and newsstands about three days before the publication date. The Houston library received their copy of that Newsweek on Friday May 25, 1951. The library was just a mile from the location of MDAH at that time, and Grimmett would have been very interested in reading that particular copy of Newsweek. The front cover showed a picture of a Tower of London Beefeater standing in front of Tower Bridge, London, with the caption, "Festival of Britain: Bright Note in a Dark Europe." This was Britain's world's fair six years after the World War II to show that it had recovered and that its products and services were

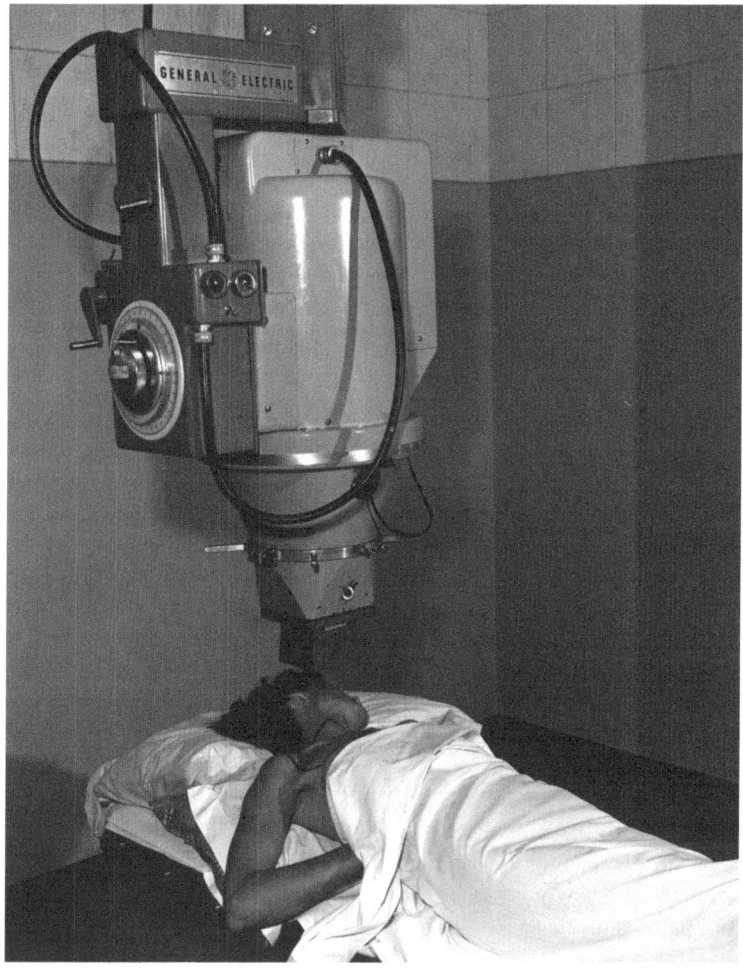

Fig. 8.5 The M. D. Anderson Hospital cobalt-60 unit in use in the hospital [139]

available for purchase by the world. It was a major event in Britain at the time and attracted thousands of visitors to London and to a rebuilt South Bank on the Thames River and the new Festival Concert Hall. Grimmett would have wanted to read all about it to see how his country was doing and also because his wife would soon be in England and would have the opportunity to visit the Festival. If he had read Newsweek that weekend, he surely would have seen the article about the cobalt unit and that would have distressed him greatly. However, none of those close to him, at that time, recall that he mentioned it and doubted that he had seen it.

But Dale Trout did read the article on the Friday before Grimmett died. He was on the way to the Cleveland airport and picked up a copy of the news magazine on his way. He immediately recognized the error and wrote Grimmett a letter on the airline stationary expressing his regret over it. Grimmett never saw the letter. In a letter to Clark, Trout apologizes sincerely for the mistake; he insisted that it was not intentional and that, in the future, G. E.'s Newsbureau would clear all releases about the unit through MDAH and ORINS. He had the highest regard for Grimmett.

> "You see," he wrote, "Leonard and I had had correspondence for a period of years before he came to this country. We had many mutual acquaintances and the British Hospital Physicists' Association, of which he was an outstanding member, was good enough to elect me to membership a few years ago. He was a classmate of one of the fellows in our London Office.
>
> In closing, let me say again, that all of us here held him in high regard and would do nothing to detract from the credit due him for his excellent work in the field of radiation physics" [138].

On Saturday May 26, 1951, Dr. Grimmett and his houseguests, Dr. Stella Booth and her husband, had a quiet dinner at home and retired to bed fairly early. By this time, Grimmetts had moved out of the rented house on Kipling Street and had bought a house at 3,238 Ewing Street a mile or so east of the site for the new hospital in the Texas Medical Center. A birthday party picnic for Grimmett's secretary, Trudy Kocian, was planned for the next day. Shortly after 1 o'clock in the night, Stella Booth heard a loud thump, on investigating she discovered Grimmett slumped at the top of the stairs, he was dead by the time she reached him. She called in Dr. Clifton D. Howe who was head of the department of medicine. He signed the death certificate and gave the cause of death as a heart attack.

Grimmett never got to see his unit in use. The ORINS/MDAH source was finally released from Chalk River to Oak Ridge in July 1952 where the unit underwent further testing for another 14 months. By September of 1953, the construction of the new hospital in Houston was far enough along that the cobalt unit was finally shipped to Houston. By February 1954, patient treatments began on the unit a few weeks before the new hospital formally opened. It was used clinically at M.D. Anderson Cancer Center until 1963 and was eventually loaded with a 2,000 Ci source, and the treatment distance extended to 75 cm (Fig. 8.5).

Chapter 9
Medical Physicist Part II, Houston, 1949–1951

Although Dr. Grimmett is primarily remembered for his work on the cobalt-60 unit, perhaps his greatest contribution was laying the foundation for a strong and viable physics department at M.D. Anderson Hospital. The mission of the institution at that time was patient care, research and education as they related to cancer, and Grimmett moved to make sure that all elements were undertaken in the new department.

He also understood that there were some activities that fell outside these broad categories with which the physics department would have to be concerned, primarily activities associated with running a radiation safety program for the hospital and which could be classified as administrative activities.

One of Grimmett's laboratory notebooks has survived and gives us a good insight as to his activities other than the work with the cobalt-60 unit [140, 141]. There are entries on 45 different dates, both on week days and weekends, from July 24, 1949, to May 17, 1951, 10 days before he died. Nearly 25 different problems were investigated in a wide range of area that today would come under the heading of nuclear medicine, dosimetry, brachytherapy, radiobiology, radiation protection, optics, electronics and general physics. The entries in the notebook are in a clear neat handwriting with meticulous drawings of the apparatus. Graphs and radiographs are taped into the notebook, and the calculations are carefully worked through. Surprisingly, there were none in diagnostic radiology.

Some of these projects are described below along with others that are not specifically in the notebook but for which additional documentation exists.

There was much to be done and Grimmett was a very busy man.

The establishment of a first-rate machine shop has already been described and was always central in Grimmett's thinking. He knew that successful radiation therapy at that time required the services of a machine shop. Patient care required the making of customized shielding blocks and accessories to maximize the effectiveness of the treatments. Research also required the ability to make unique pieces of equipment for the experiments associated with developing new approaches to patient treatment, the measurement of radiation and the investigations into the biological effects of ionizing radiation.

In August 1949, Grimmett began to expand upon his "Provisional 1949 Work Plan for the Physics Section" that he had written in February. The physic's section was now a department, and he outlined a number of programs that he wanted to undertake.

Education

The first one was a "Proposals for the Educational Progamme (sic) of the Physics Department" submitted on August 9, 1949. He proposed offering three courses in radiation physics and radiobiology [142]:

> Course (a) for postgraduate medical students intending to specialize in radiotherapy or isotope work.
> Course (b) for postgraduate physics students who wish to enter the field of Radiation Physics.
> Course (c) optional refresher course in elementary mathematics and physics.
> The courses will consist of lectures and laboratory exercises designed to familiarize the students with the practical side of their chosen subjects.

Grimmett realized that such training was vital to the radiotherapists but also would emphasize the need for trained physicists and so help establish medical physics as a profession:

> The first two educational activities are complimentary: Radiotherapists receiving course (a) will feel the need of trained Radiation Physicists to assist them in their routine and research work.

The subsequent educational activities of the physics and radiation oncology departments at M. D. Anderson Cancer Center have proved how correct Grimmett was in this assumption. Quite a few staff and residents in radiation oncology have left the institution over the years taking members of the physics staff with them.

Grimmett undertook to give the initial offering of Course (a) in the late spring of 1950 and involved both lectures and practical instruction. External beam physics and radioisotopes were covered. It must have been a success because Clark commended him upon it, and Grimmett replied to Clark on June 7, 1950:

> A report will be sent to you when it is written. The lectures were given without script, but many of those present have asked for a permanent record, and I have undertaken to write them up within a period of three months. [143]

If he did "write them up", none have survived. His death prevented him from giving the course again, which was planned for June 18–23, 1951. This course, started by Dr. Grimmett, has over the years evolved into many different courses and formats and has been continuously taught since it was inaugurated. He also gave frequent lectures to the students of the University of Texas medical and dental Schools and at the Baylor Medical College.

Before Dr. Grimmett arrived in Houston, Tom W. Bonner, chairman of the physics department at Rice Institute, had been appointed as a consultant in physics to start a program whereby physics graduate students at Rice could be accepted as fellows in physics at M.D. Anderson Hospital. These students were assigned additional research projects at the hospital, for which they were paid part time, in addition to their dissertation research. Dr. Bonner fully appreciated Dr. Grimmett's stature in the medical physics world and continued to fully support this program. There has always been and continues to be a strong link between the two physics departments.

Equipment

He next proposed, on August 12, 1949, the construction of an automatic dose contour-plotting machine [144].

Dr. Kemp in London had designed such a machine, and Grimmett wished to build one using American components. At the time of the proposal, it was primarily intended for use with the ortho-voltage X-ray therapy units and Grimmett believed that it would help extend X-ray tube life. Apparently, the General Electric Company in England owned the rights to the design, and Grimmett had already negotiated with them for the drawings of the unit, which he expected to receive shortly. He estimated the cost at $10,000 which was a sizable fraction of his budget for the coming fiscal year which was a little over $100,000 total. He was planning for the Instrument Division of the Kelley-Koett manufacturing Co., Covington, Kentucky to build the instrument having met and talked with Mr. Rasmussen of that company in early September, 1949, in Washington. This resulted in Grimmett being offered a job with the company, which he declined. Clark must have been approached also about Grimmett becoming a consultant for the company, which was news to Grimmett. "The question of my acting as a consultant for this firm was never raised" he wrote Clark on September 21 [145]. Clark was not opposed to the idea and wrote Rasmussen on September 28:

> I would be pleased to discuss with Doctor Painter, the president of the University, regarding Doctor Grimmett furnishing you with consultative advice on design and planning for our instruments, to be used in dosimetry and related physical problems in the field of radiotherapy. If any new instruments should result, we are in a position to give them a thorough clinical trial and report to you the results obtained. [146]

This was the first proposal for the physics department to work with an outside company in the development and testing of instruments and equipment. It would not be the last, and the department has a long-standing tradition for such arrangements.

Grimmett sent the drawings to Mr. Rasmussen for review in early January 1950 but heard nothing back. The delay was due to Mr. Rasmussen being on an extended trip and in the end the cooperation with Kelly-Koett did not materialize, but the project to build an automatic isodose plotter, continued in the department.

Mention has already been made of Dr. Spurr and the beginning of the radio-isotope program at the M. D. Anderson Hospital. Spurr had ambitious plans including a central laboratory for handling lager amounts of radioactive isotopes, a laboratory for detection and counting of the isotopes especially tracer amounts, storage and disposal facilities for radioactive materials such as urine and blood and the necessary equipment to carry this out. Various research projects would also be undertaken. One month after Grimmett had arrived in Houston Dr. Spurr was involving Grimmett on these projects. However, Dr. Spurr left the institution in July shortly after Grimmett arrived and with so much else to be done the radioactive isotope program only slowly moved ahead. Just five days before he died, Grimmett addressed this problem again with Clark.

> When we talked about the isotope program last week, you asked me to send you a reminder about the extra physicist-technician we discussed...I for my part would enthusiastically welcome an opportunity of bringing our isotope work out in front, by putting into operation all the various ideas which have been mooted in the past, but which have never been implemented for lack of help. We have located such a man at the Rice Institute- Mr. Kohl, who intends to specialize in electronics and biophysics, and would be keenly interested in joining us here. [147]

In fact the man's name was Cole—Arthur Cole. Grimmett's German born secretary Trudy Kocian reverted to the German spelling in typing the memo. Arthur Cole did join the institution, earned a Ph.D in biophysics and stayed with the institution until his retirement. He was also gifted in electronics and built the automatic iso-dose plotter for the department. Unfortunately, Grimmett's untimely death prevented him from seeing its completion.

The Allis-Chalmers 22-Mev Betatron

M.D. Anderson was the second hospital (after Memorial Hospital in New York City) in the country to install an Allis-Chalmers 22-Mev betatron. The hospital was clearly committed to getting such a machine and by November 1950 Grimmett was doing experiments on the transmission of light through sheets of thick plate glass that he was considering for the viewing window into the proposed betatron room. He was apparently working on the plans for the shielded rooms in the new hospital, and he was pushing Clark to make a decision about placing an order with Allis-Chalmers for the betatron. In December of 1950, he sent a memo to Clark stating:

> It would seem that there is likely to be a great delay in getting a betatron unless we should be in a position to place an order soon. [148]

Clark replied that he "Would like nothing better than to place orders for all equipment for the new building now." [148] But it was not possible until the next legislative appropriations in early 1951. In the meantime, he suggested a priority list be readied to go when the money became available. On his April 1951 trip to G. E. in Milwaukee, Grimmett met a representative from a glass company in Seattle who had samples of a new lead class that was amber-tinted and which contained 55 %

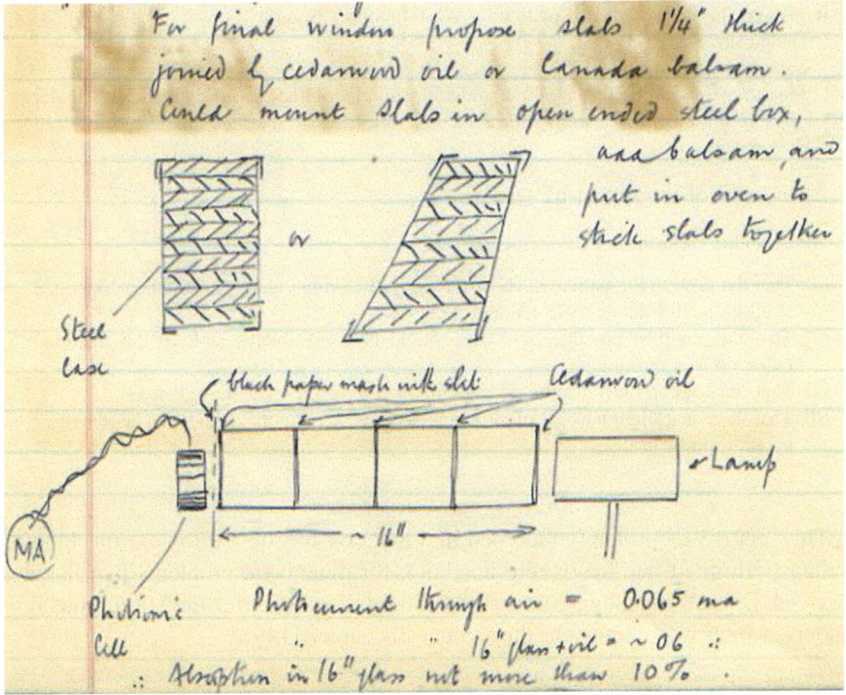

Fig. 9.1 Entry from the notebook on November 10, 1950, concerning the glass for the viewing windows

by weight of lead. They were 4" thick (the samples he had previously measured were only available in 1¼" thickness) and were "… of excellent optical quality." [128] He immediately recognized that this would be ideal for the viewing windows for both the cobalt and betatron treatment rooms in the new hospital, and he made a note to bring it to the hospital architect's attention (Fig. 9.1).

The hospital did place an order for the betatron but it too, like the cobalt unit, had to wait for the new hospital to be ready and was installed in November 1953. The viewing windows for both treatment rooms used the glass Grimmett had seen in Milwaukee.

Diagnostic Radiology and Protection

It is not surprising that there are no entries in the notebook dealing with diagnostic radiology, since the hospital administration saw physics as mainly supporting radiotherapy This does not mean that Grimmett was not involved in diagnostic radiology problems, especially if it involved radioprotection, since the Radiology Department combined both diagnostic and therapy applications.

At least one diagnostic report has survived:

Report on the measurements made on a Westinghouse 'Autoplex'
Diagnostic Set.
Saturday, August 26th, 1950
Measurements were taken of the radiation output of the fluoroscopy tube... under various conditions, and also of the protection afforded by the machine and some lead rubber aprons.

At the end of the report he wrote:

We conclude from these measurements:

(a) That the dose to the patient is about the same, 12 r/min., whether the tube is operated at 90 kV and 3 ma or at 75 kV and 5 ma.
(b) That the smallest possible field size should be used always to keep the scattered radiation to a minimum.
(c) That the lead apron transmits not more than 3 % of the incident radiation.
(d) That there is a slight leakage of radiation from the X-ray cone, which can be remedied by addition of lead.

This report was probably recorded in a logbook for the 'Aotoplex' machine or perhaps Grimmett had a separate notebook for diagnostic problems that has not survived. But it does show Grimmett's continuing concern for radiation protection matters, which was a constant throughout his career [149].

Dosimeters

Ionization Chambers

Grimmett also had a continuing interest in the development of dosimeters of all kinds. He had planned to mold his own ionization chambers in Houston using an air equivalent plastic that he had developed in England before the war, which would make the ion chamber response independent of the X-ray energy. He was very impressed by Sievert's chambers that were constructed of "Electron" metal (mainly manganese with small amounts of aluminum, zinc and copper and was not energy independent) and had wanted to get a molding machine so that he could make his own chambers using his plastic (a mixture of bakelite and graphite with a small amount of titanium or vanadium oxide). This material could be conveniently molded under pressure, and the chambers had excellent electrical and mechanical properties.

The notebook contains several entries concerning ionization chambers from using them in experiments to calibrating them. In addition to the Sievert chambers, the department also had a Victoreen R-Meter with five different chambers. Grimmett was interested in determining the exact center of the air volume for these chambers and took radiographs of all five and then made full-scale diagrams from the radiographs as shown in Fig. 9.2. This was the chamber he used in his radiobilogical experiments.

Dosimeters

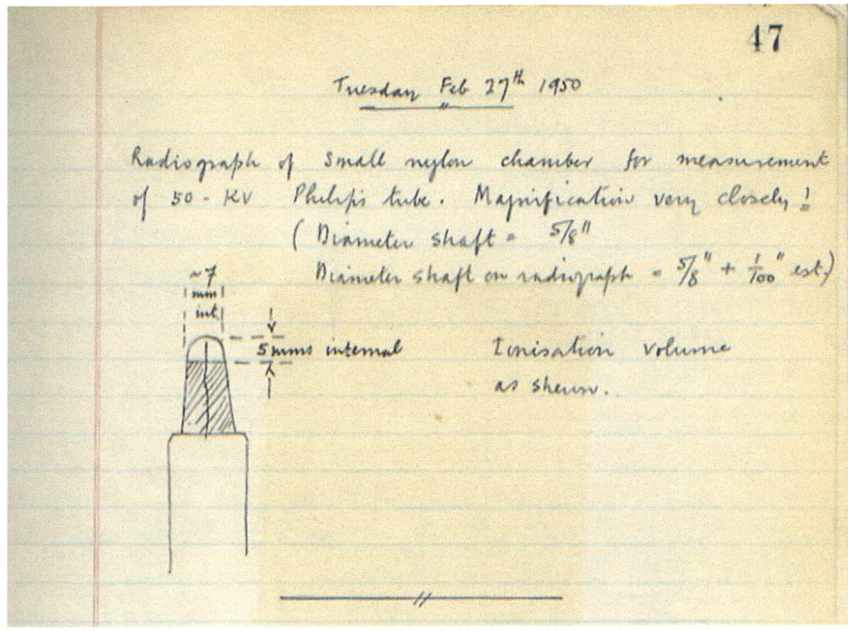

Fig. 9.2 Entry from February 27, 1950, concerning a small nylon Victoreen chamber. A radiograph of the chamber was taped to the opposite page of the notebook

Scintillation Detectors

Grimmett and Jasper Richardson worked on scintillation counters in 1950, and Grimmett presented a paper on their work to the Southwestern Section of the American Association for Cancer Research meeting in Austin in December 1950.

Some of the work is recorded in Grimmett's notebook for September 25, 1950, under the heading: "Notes on $CaWO_4$ crystal". This was probably the crystal that he reported on in the paper, although it does not explicitly say so. He used the gamma rays from I^{131} to test the crystal and calculated the absorption coefficient for the gamma rays in the calcium tungstate. He had three crystals 1.0, 2.0, and 3.8 cm long, respectively, and he also studied the "Efficiency of counting in long crystal in various direction." Either perpendicular or end-on to the gamma rays and concluded that "...the end-on position will have smaller count due to γ absorption."

The paper was entitled "Notes on the use of scintillation counters in radiotherapy". He sent a copy of the paper to Richardson at the beginning of 1951 stating that:

Unfortunately, it will not be published. [133]

Apparently, they were trying to develop a very small detector to measure leakage radiation along the joints between two metal blocks, probably in

Fig. 9.3 Grimmett scintillation probe positioned near the experimental stainless steel ovoids (The scintillation counter was further developed by Shalek and Cole and reported in the early papers on the stainless steel ovoids. Radiology Vol.60 No.1 pp. 83–84, 1953) [150]

preparation for measurement on the new cobalt unit, but their counter did not prove useful for this since the sensitivity was too low. However, it was used to measure the dose distribution around a 1-mg radium source and the isodose curves around the stainless steel ovoids that Fletcher and Grimmett were developing (Fig. 9.3).

Film Dosimetry

Grimmett was also very interested in developing a wavelength (energy)-independent film. He had first proposed this to the administration in December 1949 and by mid-1950 he had revised the proposal and resubmitted it to the hospital administration. He was proposing to work closely with the ANSCO film company which was working on a similar program. Grimmett pointed out that the normal variation in film sensitivity between 50 kV and 2 MeV was a factor of about 16. ANSCO had already reduced this factor down to about 2, and Grimmett believed it could be reduced even further but the research would be costly, and ANSCO suggested that they would need $15,000 to $25,000 of outside funding to embark on such a project. Although costly Grimmett thought it was well worthwhile pursuing. He made the further suggestion:

> It occurs to me that, if the money can be found, and if ANSCO is willing to undertake the project, then in the event of a successful product it might be desirable to have the name Anderson Hospital attached in some way, in order that some credit may come to the institution for the stimulation and financing of such research, e.g. we might persuade the firm to call the product 'Ansco-Anderson film' or some such device which will defer to its origin. [151]

The money, however, could not be found and the project was dropped.

Chemical Dosimetry

In September 1949, he commenced a study on chemical dosimetry. The heading in the notebook states:

Wednesday September 27, 1949
   Chemical Dosimetry
   (Rough preliminary experiments)

His first attempted was at colorimetry, using the change in color of a solution with dose, as measured by optical density, as a dosimeter. He used water saturated with benzene and developed with Folin-Ciocalteu, a reagent for phenol. George Awapara, who had helped Grimmett settle in when he first arrived in Houston consulted with him on this project and he used Awapara's colorimeter for the project. Grimmett recorded that:

> Colorimeter setting 4700 A° for this reagent. The colour changes were not visible to the eye[*] There is a linear relationship between the dose and colour density.
>   * Awapara says colour fades quickly. Readings on density must be taken immediately after adding Folin-Ciocalteu.

Indeed there was a very good linear relationship, although he only measured three points.

A few weeks later on in October 1949, he continued his studies on chemical dosimetry; this time with solutions of ferrous sulfate ($FeSO_4$) and ferric chloride ($FeCl_3$) using dimethylglyoxime as the reagent. He recognized the need for fresh solutions, and the need for pure water. "Should repeat with freshly made solution-with boiled distilled water (Presence of O_2 would give H_2O_2-oxidising agent working against reduction process.)" This time he used four dose levels for both the ferrous and ferric solutions but found poor correlation with dose.

On Monday October 24, 1949, he wrote:

Conclusion. Fe reaction not much good for dosimetry (Fig. 9.4)

For once he missed the mark. Chemical dosimetry and in particular ferrous sulfate, or Fricke dosimetry after the person who first described it [152], became well understood and can be used for the precision measurement of radiation dose. Later on Grimmett's department made extensive use of the dosimeter [153], and several national calibration laboratories have used it in their work.

Although Grimmett had the help of his good friend George Awapara, who was a biochemist, it is clear from the notebook that he also had a clear understanding of the chemistry involved.

He returned briefly to this subject in October 1950 under the title, "Chemical Dose Indicator". He used chloroform and brown cresol purple in water above the chloroform in a 5-ml flask and irradiated the flask to 250-kV X-rays. There was no change in color up to 334r when the brown cresol purple turned burgundy red and at 648r, it became a bright yellow. "Nucleonics" is written on the page and a note which said, "The water layer above contained brown cresol purple adjusted to pH as in above article." The exact reference was not given.

Excretion of I^{131}

It was not noted in Chap. 7 that on February 15, 1949, Grimmett submitted his "Provisional 1949 Work Plan for the Physics Section" and had added a hand written note stating that… "the Physics Department could…cooperate with the Clinical Department in the administration…of radioactive substances…" It should be no surprise therefore that the first subject he tackled in the notebook on Tuesday July 26, 1949, concerned radioactive I^{131}. Neither should it be a surprise that the first entry was five months after submitting his work plan, considering all the other things he had going on during that time period.

The entry reads:

July 26, 1949
  Problem
  Excretion of I^{131}
  Given $Q = Q_f(1-e^{-rt})$, to find Q_f and roentgen (r) from observed values of Q and t…

This was a familiar problem of the day when thyroid problems were studied, either clinically or in research animals using I^{131}, by measuring the amount of iodine excreted in urine [154]. "t" was time and Q would have been the activity in the urine.

His first approach was to record the counts Q_1 and Q_2 at times t and 2t, and the solution to the equation required solving a quadratic equation for r.

"Another method" he looked at was to get values of Q at three equally spaced values of t and to derive an approximate value of Q_f.

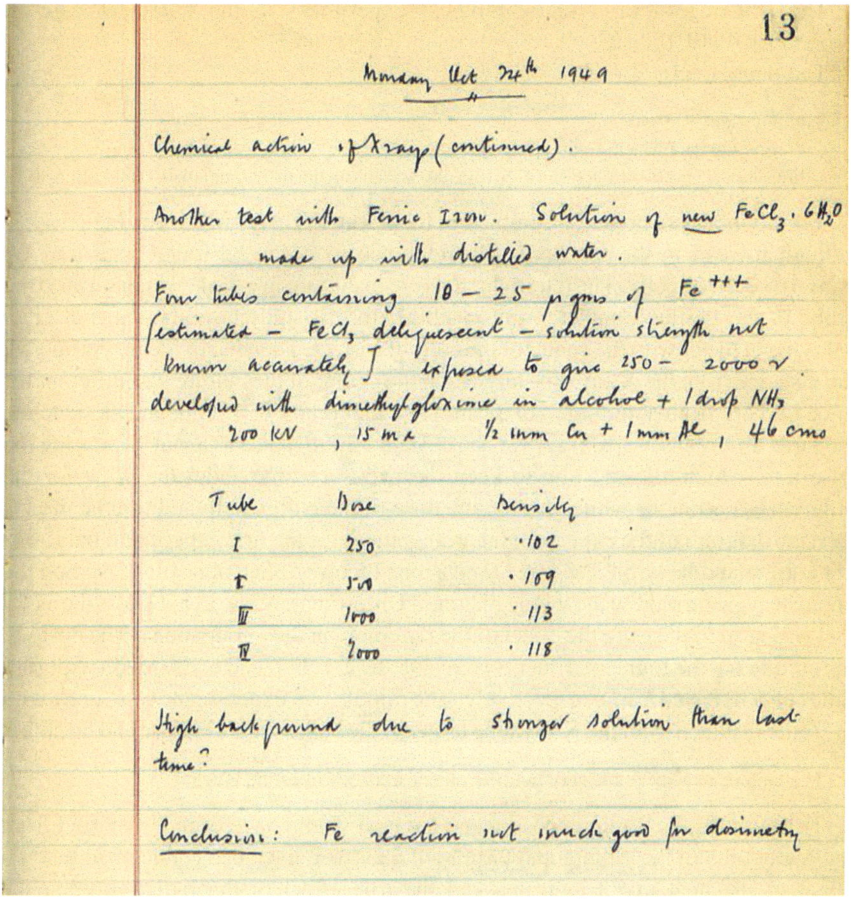

13

Monday Oct 24th 1949

Chemical action of X rays (continued).

Another test with Ferric Iron. Solution of new $FeCl_3 \cdot 6H_2O$ made up with distilled water.

Four tubes containing 10 – 25 μ gms of Fe^{+++} (estimated – $FeCl_3$ deliquescent – solution strength not known accurately) exposed to give 250 – 2000 r. developed with dimethylglyoxime in alcohol + 1 drop NH_3

200 kV, 15 mA ½ mm Cu + 1 mm Al, 46 cms

Tube	Dose	Density
I	250	·102
II	500	·169
III	1000	·113
IV	2000	·118

High background due to stronger solution than last time?

Conclusion: Fe reaction not much good for dosimetry

Fig. 9.4 Notes on Chemical action of X-rays from Grimmett's notebook October 24, 1949

Finally he solved to problem graphically for Q_f.

He was satisfied that all three methods gave the same result.

No actual data was involved, this was purely an exercise in mathematics and shows Grimmett's skill as a mathematician. Today this is a trivial problem readily solved on a hand held calculator.

Radiobiology

It was seen in Chap. 4 that Grimmett was interested in radiobiology and was a joint recipient in 1934 of a prize from the British Empire Cancer Campaign for work on the action of ionizing radiation upon malignant cells. This interest remained throughout his life.

The first record of a radiobiological experiment in the notebook was on September 9, 1950.

The entry reads:

Saturday Sept 9, 1950
  Philips Contact Machine 50 kV 2 ma
  Preliminary expts. On action of X rays on Tetrahymena Browningiensis (sic)

Tetrahymena are protozoa common in fresh water. The word 'Browningiensis' is difficult to read in the notebook. No reference to it can be found so it may be transcribed incorrectly. The tetrahymena were in culture in a flask, which sat atop the probe of the Philips Contact X-ray machine. In this study Grimmett was at great pains to get the dose to the center of the culture correct. There were four experiments, the distance from the X-ray tube target and the center of the culture was different for three of them from a minimum of 22 mm to a maximum of 45 mm and he sketched the three different set-ups. It appears he changed the distance as a way of varying the dose rate, which allowed him to keep the exposure time about the same for the different experiments while varying the dose to the cultures. In addition he drew a full size diagram of the experimental arrangement for the first experiment but it was not correct and he wrote: "Some Dimensions Incorrect. See opposite." And on the opposite page he redrew the whole diagram. Unfortunately we do not know the assay that was used to measure the effect of the radiation on the tetrahymena. The doses to the culture for the four experiments were 498, 672, 1,344, and 6,700 r with exposure times of 4, 4, 8, and 5 min respectively. The middle two experiments were done with the same distance from the X-ray tube target to the center of the culture and he noted:

> The culture in Expt 3 got very hot; the others were warm to the touch.

This was for the 8 min exposure and the heat would have been transferred from the X-ray tube to the culture and Grimmett may well have been concerned that the effect of the heat might have masked the true effect of the radiation upon the culture. This might explain why he varied the dose by changing the distance from the target rather than just using longer exposure times.

Clinical Research

Breast Treatments

In light of the subsequent rift between Fletcher and Grimmett it is interesting to note that Grimmett was directly involved in clinical research. A set of data sheets has survived in the Department of Physics archives for dose measurements on patients being treated with 250 kV X-rays for breast cancer. Between October 6 and November 28, 1949, 22 sets of data were taken on three patients. Six Sievert chambers were placed on the treated breast. The data sheets record the treatment parameters, the position of the chambers on a diagram and the dose recorded by each chamber. Grimmett was personally involved and signed many of the data sheets and a notation was made that a copy of the data sheet was sent to the Radiotherapy Department. In addition a phantom study was done using a press-wood phantom the "Field size, chamber arrangement, beam direction and exposure time were the same as on patient..." Recorded on the data sheet is the object of the experiment, "... to show that under ideal conditions the dosage of the chambers would be approximately alike." This sheet was also sent to the Radiotherapy Department. What all this data was used for is not known but it is clear that Grimmett and Fletcher must have worked closely on this.

Skin Reaction

One of the largest projects in Grimmett's notebook was what he called the "Wax Block Experiment". He described it as:

> Preliminary measurements for G.H.F's (Fletcher's) experiments on human skin reaction, with and without a wax block on the skin; he suspects increased skin reaction due to soft radiation under wax block, *for equivalent number of roentgens as measured by a conventional ionization chamber.* Problem is to find the absorption in a wax block, in order that the exposure time may be increased to compensate the diminished dose rate.

This project occupies pages 20 through 39 of the note book and went from Sunday January 22, 1950, to Saturday February 4, 1950, with most of the work being done on weekends.

Wax blocks were placed on the skin of patients to compensate for variations in the skin surface, resulting in a more uniform dose distribution at the depth of the tumor. It was often used to compensate for the slope of the breast and this may have been Fletcher's interest here. In all Grimmett did four different experiments.

It appears that Fletcher was concerned about the skin reaction, always a problem with orthovotage X-rays, and not helped at all by the wax blocks. Grimmett found that the dose at the interface between the skin surface and the wax block was diminished by 14 % but,

> To compensate the dimunation of dosage rate the exposure time should be increased... by 16 %

Grimmett realized that his chamber walls were too thick and commented that:

> The foregoing measurements should be supplemented by measurements with very thin-walled flat ionization chamber, suitable for the detection of soft radiations. Is chemical reaction of radiation products in wax an alternative explanation of the suspected increased skin reaction?

It was the electrons coming off the face of the wax block that was the problem. As it turned out it would be the cobalt unit that resolved this problem. Metal lead wedges could be used to compensate for the slope of the breast and variations in the patient's skin surface could be accounted for with metal compensators that could be scaled down in size and placed in the head of the cobalt unit away from the skin surface (about 15 cm was required) and any electrons coming off the wedge or compensator were absorbed in the air before reaching the patient's skin so preserving skin sparing.

Cervical Cancer Applicators

Grimmett was very much involved with the design and development of the MDAH stainless steel shielded applicators for the treatment of cervical cancer. This is not generally recognized since the initial publications on them did not appear until some time after his death and without mention of his name.

The American Cancer Society field notice 4-30-48 had advised medical institutions to delay further purchases of radium until the availability and suitability of using cobalt-60 as a radium substitute had been determined and Fletcher was interested in pursuing this in a number of directions. He wanted to see if cobalt-60 needles could be developed to replace the radium ones for use in interstitial implants and also if a technique called the radium "cosine law" surface applicators, proposed by the South African J. van Roojen from the Department of Radiology at the University of Cape Town, could also be converted to cobalt [155]. And of course he wanted to investigate if cobalt-60 could be used in teletherapy units. On his first visit to Oak Ridge Institute for Nuclear Studies in August–September 1949, Grimmett lectured on all these proposals.

But Fletcher and Grimmett were interested in another application, the development of applicators for treatment of cancer of the cervix using cobalt-60. In the research program of the radiotherapy department that was sent to Clark in August 1949 Fletcher wrote:

> Many radium applicators have been developed which use variations of either the so-called Stockholm or Paris techniques. In 1944, Neary, through very careful calculations, showed that some of the principles which had guided the design and anatomical positioning of the applicators were wrong and showed by his calculations that it was possible to increase the dosage to the parametrium without undue increase of dosage to the vaginal wall, rectal septum, bladder and pericervical tissues. Such an applicator has already been in use at the Mt. Vernon Hospital in London but still further clinical work has to be done. It involves a large amount of radium and platinum but these materials could be replaced by cobalt-60 and metallic uranium. [141]

As a preliminary study it was proposed to construct an applicator of the kind developed in Manchester (England) with a thick sheet of gold on the rectal side for shielding. The Manchester applicators were made out of a hard plastic and were approximately egg-shaped and were called "ovoids". Measurements around the applicator with very small ionization chambers and rechecked on patients with the chambers placed in the rectum and bladder was also proposed. The ion chambers were the Sievert chambers that had been obtained from Stockholm and which had just been calibrated at the National Bureau of Standards (NBS). At some point it was decided to make an ovoid out of stainless steel, which reduced the size and made the shape more cylindrical and tungsten was used for shielding. Uranium was almost impossible to come by and Grimmett had extensive experience using tungsten. Grimmett continued to call them ovoids but the name they came to be known by was "colpostats", although eventually that too was dropped and they are now generally known by the old designation of "applicators". Suitable cobalt-60 sources did not materialize and the stainless steel ovoids where developed for use with radium. Grimmett undertook the first dosimetric study of them in early April 1950. He measured the dose distribution around both stainless steel and plastic ovoids, with and without shielding [140]. The dose rate was about ten percent lower with the stainless steel ovoids but this could be accounted for by the treatment time, the actual dose distributions were very similar.

Treatment Machine Calibrations

His major concern, however, during this time was the calibration and maintenance of the clinical X-ray machines. This would eventually become a major source of contention between himself and Dr. Fletcher. Dr. Fletcher was convinced that there were serious errors in the calibration of the various orthovoltage X-ray machines at the hospital and he came to believe that Grimmett was not doing enough to correct the situation. In June 1949, Fletcher sent a detailed memo to Clark pointing out the problems they were having with calibrating the X-ray units [156]. The main calibrating system in use at the time was the Victoreen dosimeter consisting of a string electrometer and a number of Victoreen condenser chambers. To have them calibrated required sending the dosimeter back to the factory in Cleveland for calibration against a standard chamber. This took time (about three weeks) and in the latest calibration the string in the electrometer broke when it was shipped back to the hospital and the process had to be repeated, and Fletcher was not too happy. He was writing Clark to get permission for one of the physicists at the hospital to go to the NBS in Washington D.C. hand carrying a Farmer electrometer and a number of Sievert condenser chambers for calibration.

Fletcher concluded his memo by saying:

> It cannot be emphasized enough how important it is to have constant and accurate checking of the output of the machines...
> The construction of a standard chamber should be considered seriously as part of the Physics Department for use when we are in the permanent hospital.

Grimmett tackled the problem head-on. He wrote the suppliers of the Sievert chambers in Stockholm, whom he personally knew, to get the sensitivity in "volts per roentgen" for the chambers they had supplied [157]. He wrote Dr. Lauritson Taylor at the NBS to make arrangements to calibrate the chambers and scheduled August 15 to be at NBS in Washington D.C. to carry out the measurements [158]. He wrote his friend W. Binks at the national Physical Laboratory in London (the British equivalent to the NBS) asking for his help:

If you were faced with the problem,

(a) what design of chamber would you go for (say for 50 kV to 400 kV?)
(b) what sort of X-ray machine would you ask for?
(c) what special facilities would you consider essential? (e.g. is a constant voltage transformer necessary, what design of stabilization is required on the chamber voltage, etc.)

I feel that a few hints from you might save me an awful lot of mistakes, and set me on the right road from the beginning. [159]

He found out from the North American Philips Company that they had their ion chambers calibrated by Carl Braestrup in New York and he made arrangements through the Cranford X-Ray Company in Houston to have his Victoreen electrometer and chambers calibrated by Braestrup in late December 1949. All of this was done at a time when he was beginning to become deeply immersed in the cobaly-60 project.

It is also clear that questions about who would control the activities of certain personnel in the physics department were surfacing. The institute had an opening for a radiation physicist and Grimmett had been in correspondence with Mr. Peter Wootton in England about the position. His qualifications were good and there was general agreement that he should be offered the position. In late January 1951, the administration held a conference with Fletcher about various personnel issues in the institution. Grimmett was not present and when Wootton's appointment was discussed Fletcher insisted that:

...his appointment be primarily in clinical radiation physics and that he should be assigned in all his routine work to the department of radiation therapy under his (Fletcher's) direct supervision. However, he would also have an appointment in the Department of Physics where as assistant physicist he would work with Doctor Grimmett in performing his research in radiation therapy. [160]

One can only imagine Grimmett's distress when he heard of Fletcher's conditions for hiring Wootton, especially since Grimmett had carried out all the negotiations. By April 1950, the situation had reached a critical point.

Grimmett had gone on a business trip ("on mission" as he put it) and in his absence Fletcher took matters into his own hands. He asked Mr. McLean, Grimmett's assistant physicist, to calibrate the 200 kV machine. It was generally considered that X-ray machines operating under the same conditions of voltage (KV), current (MA), focus to surface distance (FSD) and filtration would have about the same output in terms of roentgen (r)/min. The measurements were done on Saturday April 1st with the machine's parameters set at 150 kV, 15 MA, no filter. On Tuesday April 4, 1950, he sent a memo to Grimmett on the "Calibration of Therapy Machines."

In part he wrote:

The result was 90 r/min at 50 FSD. Previous measurements made on the 12th of February using the same factors (150 kV, 15 MA, no filter), but on the 250 kV machine showed 58 r/min.

Although these are two different machines, I feel that these two measurements show too much of a discrepancy and should be rechecked before treatment is instituted.

He went on to point out that he had patients waiting to be treated and he asked for the Victoreen dosimeter so that his department could do the calibrations and suggested that as a matter of routine they do all the machine calibrations since,

> ...so far it as been impossible to get the routine calibration at regular intervals... It would be greatly appreciated if both advice and equipment could be given by the Physics Department in order that the Victoreen can be kept as dry as possible. [161]

Grimmett received this memo on his return to the institution and he did not take it lying down. He immediately calibrated the two machines in question himself and then fired off a two-page memo back to Fletcher. The results of his calibrations were not much different than the values quoted by Fletcher. Grimmett measured 89.6 r/min (compared to 90 r/min) on the 200 kV machine, and 67.5 r/min (compared to 58 r/min) on the 250 kV.

> I do not feel that the differences between his (McLean's) readings and my own have much significance, in view of the many causes of error-fluctuating voltage, poor localizers, inconstant dosemeters (sic)- causes which I have pointed out <u>ad nauseam</u> (Grimmett's underline).
>
> I should be very willing to keep the Victoreen in good condition so that you may calibrate the X-ray tubes yourself.
>
> You ask for my advice. The advice I would give is that you would not delude yourself into thinking that the use of a Victoreen chamber under present conditions will bring accuracy into your work.

Grimmett then went on to point out what the problems were. Number one in his mind was the instability of the voltage supply and he attached some of his measurements showing the variation of output versus line voltage. He pointed out that equipment that he had asked for had not been ordered, including a stabilizer for the voltage supply. Secondly was the need for a good localizer to ensure accurate positioning of the chambers for calibration, He had planned to build a new one of his own design:

> Personally, I would have given this high priority in the workshop, but since you preferred to have the experimental ovoids and Heyman applicators made first, I have deferred to your wishes.

Thirdly was the lack of good measuring instruments and standards. He discussed at some length the situation of the calibration ionization chambers and the need for the Institution's own a standard chamber.

> Plans are ready and waiting, but with the long lists of clinical requirements confronting the workshop, it has receded very much into the background.

In conclusion Grimmett wrote:

> Lastly, may I take this opportunity of requesting that you will not interfere with my staff when I go on mission? Both Mr. McLean and Mr. Mutrux have complained. It was my understanding that they would work under my direction. This sort of recrimination seems to be uncalled for, and shows a lack of appreciation of the difficulties we are struggling with. I have the dosimetery problems very much at heart, and have pointed out very clearly what needs to be done to put the whole matter on a proper footing. But I object very strongly to wasting my time, and that of my staff, in the accumulation of worthless data. [162]

For good measure he sent a copy to Clark. This is a wonderful example of British reserve, understatement, condescension and sarcasm. He started out the memo seeming to agree with Fletcher and offering to assist him in calibrating his treatment machines knowing full well that Fletcher was not about to do that and in any event he, under no circumstances, would let it happen. Then he systematically destroys Fletcher's charge that he, Grimmett, is responsible for the uncertainties in the calibrations and turns it around and says that, in fact, it is Fletcher who is to blame. By the end of the memo he can hardly contain his anger and tells Fletcher he will not tolerate any interference from him in running his department, and what Fletcher has done has all been a waste of time. Fletcher would not have missed the point and the relationship between the two rapidly deteriorated. At times they would not speak to each other and messages were passed back and forth by their secretaries and the situation continued to deteriorate. By September of that year, 1950, it was necessary to define the responsibilities of the two departments. In a memo sent to Fletcher with copies to Clark, Grimmett and others, Dr. Heflebower the Assistant Director wrote:

> The Section of Radio-therapy is responsible for the treatment of all patients by X-ray and Radium, and this is interpreted to include research undertaken to improve the methods in use or to find new ones.
>
> The principle role of the Department of Physics is to assist the Radio-therapy Section by making the necessary measurements and checks of dosage, etc., and the fabrication of devices which will be used in the clinical application of the X-Ray, Radium, etc. [163]

The memo was initiated because there was still uncertainty in the responsibilities between physics and radiotherapy and whose budget would pay for certain items of equipment and supplies. It is instructive to note that Dr. Heflebower saw this as a question between the Section of Radiotherapy and the Physics Department and not between physics and the Department of Radiology of which Fletcher was the chairman. The main mission of physics in the institution at that time was to work in the area of radiotherapy.

Things did not improve. At the end of April the following year, 1951, Fletcher was pushing Grimmett for the completion of the metal ovoids and measurements to be made so that clinical studies with them could be initiated. Fletcher hoped to present a paper at the Radiological Society of North America (RSNA) in December and by then he wanted, "...enough actual use of the applicator to demonstrate the advantage over the plastic ovoids as used in Manchester." [164] Perhaps Fletcher was somewhat justified in his concern. A year earlier the design of the ovoids was about complete and Grimmett had acquired a significant amount of measured data but it appears that Grimmett had not kept Fletcher fully informed.

Then in May a series of events concerning a new orthovoltage X-ray therapy machine, a G. E. 250 kV Maxitron, led the relationship between Fletcher and Grimmett to spiral out of control. Fletcher noted in a memo to Grimmett on May 2, 1951, that the Maxitron had been, "...purchased because of its great range and flexibility...but if nothing is done the expected advantages of the unit will never be fully investigated."

Fletcher had several complaints; a treatment chair that Grimmett had designed and was building in the workshop was not finished, although all the components had been sent to Grimmett months ago, and patient treatments were being compromised. Also the light localizer was unsatisfactory a fact, he said, he had pointed out to Grimmett at coffee one day but nothing had been done and the therapy nurses found the situation very difficult when taking films. In addition only a fraction of the possible filter/kilo-voltage combinations had been calibrated although he had given Mr. McLean a list of combinations to be used clinically. And finally, as far as he knew, no half-value layers had been measured.

When Grimmett received this memo he drew a heavy pencil line diagonally across it and wrote above the line "lies" in a larger and bolder script than he generally used [165].

But Fletcher was not finished. The next day, May 3, he wrote another memo to Grimmett, this one on the "Justification of the use of the share of the institutional Grant of the American Cancer Society given to combined projects-Radiology and Phyiscs."

Dr. Heflebower (the Assistant Director for Administration) had asked Fletcher for an update on the projects for which Fletcher had signed and was now responsible. The projects, Fletcher had told Helflebower that were to be undertaken by physics, included direct measurement with Sievert chambers, wedge filters, metal ovoids, and volume distribution with trans-vaginal cones. Equipment had been bought and Fletcher told Grimmett in the memo;

> All I could say was that to my knowledge the various projects had not been started and he (Dr. Heflebower) wants to know why.

Fletcher went on to point out to Grimmett that these were not of academic interest but had direct bearing on clinical care. He thought some data had been obtained with film on the cones but he couldn't find any in his files. Also the previous summer Mr. Shalek had been hired to work on the metal ovoids. He had undertaken a few weeks of measurements in September and October, 1950, but since then no more work had been done. A copy of this memo went to Dr. Heflebower. When Grimmett received the memo it got the same treatment as the previous one with the dark diagonal pencil line and "lies" written in a bolder and larger script than before [166].

Although just what Fletcher was referring to is not quite clear, what is clear is that much of the problem was due to a lack of communication between the two. Physics department records show that as early as November 1949 extensive direct (in vivo) measurements had been undertaken on post-operative breast patients with the Sievert chambers. Extensive and detailed measurements around the metal ovoids had been completed in May 1950 and measurements on the output of the trans-vaginal cones had been carried out in January 1951 and sent to Fletcher, although no record of the volume distributions have been found.

Fletcher must have thought that he did not get any satisfactory results from his two previous memos to Grimmett so on May 7 he appealed to Dr. Heflebower to help resolve the problem. He wrote:

> As I have asked you to please intercede in the matter of the checking of the accuracy of the light localizer of the Maxitron I am outlining the events in chronological order.

Fletcher listed, in detail, the problems he was having with Grimmett. The problem with the light localizer started around mid-April (apparently shortly after the machine was installed). The light beam was not coincident with the X-rays and this was causing problems for the three nurses treating the patients. Monday morning, May 7, Grimmett had called Fletcher and told him, "...that the light localizer was built wrong but there was one position of the turret that gives perfect superposition of the two beams." Fletcher asked the three nurses if they had been told what that position was and all three denied having been told. Fletcher was so upset that he called Dr. Heflebower to come to the department and witness the statements of the nurses. In conclusion he said of the actions, or more precisely the non-actions of the physics department, "This in itself is gross negligence." [167]

Wednesday evening, May 9, 1951, Grimmett, along with personnel from GE investigated the reason for the misalignment of the light and X-ray beams by taking a pinhole photograph of the machine's target and found that the localizer mount had to be repositioned, 4.3 mm toward the anode, along the anode–cathode axis and 1.7 mm in the orthogonal direction. When this was done another pinhole photograph showed the target lined up with the localizer. He also measured the half vale layer (HVL) for the Thoraeus III filter at 3.2 mm Cu, which Fletcher had complained had not been done [140]. They finished up at around 11 o'clock that evening and Grimmett left a note for the therapy nurse, Mrs. Rita Hendley, that the light localizer was O.K. [168].

A week later another problem with the Maxitron came up which added to Fletcher's distress, the output of the machine was varying more than he thought it should. This was discovered on Saturday May 12 by Charles McLean who went directly to Fletcher. McLean had previously measured the output of the unit on Saturday May 5 as 55 r/min and had also determined the HVL to be 2.8 mm Cu but on May 12 the output had jumped to 65 r/min. On the following Monday, May 14, Fletcher had waited all day for physics to come and investigate the problem. When late in the afternoon Grimmett had not shown up Fletcher asked Heflebower to call Grimmett to come and investigate the situation, but nothing was done and Fletcher fired off another memo to Grimmett, with a copy to Heflebower, saying it was imperative to get something done and that the local manager for G.E. should be brought in [169]. But nothing happened and the next day Fletcher went directly to Clark, the director of the hospital, sending him a copy of the memo he had sent to Grimmett [170].

All of this sounds like the complaints made by Constance Wood, Boag and Howard Flanders back in 1944 when Grimmett was at the Hammersmith Hospital in London. There too the complaint had been that Grimmett was never around when you needed him! It also appears that Fletcher and Grimmett were only communicating by memos or through third parties. Fletcher took the time to call Heflebower and have him ask Grimmett to deal with the problem, but Fletcher would not call him directly [171].

Memos were now going back and forth almost daily. On May 15, the day Fletcher had written Clark about the change in output, Grimmett wrote Fletcher on the subject pointing out that he and Mr. Welman from G.E. had carefully gone over the whole situation. It was their opinion that when the target was moved relative to the localizer the energy of the beam had been changed resulting in a greater output. They had measured an increase in the HVL from 2.8 to 3.2 mm Cu which could explain the output increase and he proposed to confirm this with a series of measurements the next day. In any case he thought the increase in output was permanent [172].

At some point on the same day, Fletcher wrote another memo to Clark outlining the chronological order of events surrounding the light localizer. This might have been at the request of Clark to have all the detail written down, since Fletcher gave Clark a daily and sometimes hourly account of what had happened. In any event, the problem of the light localizer appeared solved, and it also seemed to explain why the output had increased [173]. But Fletcher was not satisfied. On May 16, he sent another memo to Clark, subject, "Standard procedures in new therapy machines." In the memo, he listed five procedures which he said were standard in England when new machines are installed. Three of the five have not yet been done on the Maxitron, and these procedures, he pointed out, are to be done by the physicist and not the manufacturer of the equipment (G.E.).

He concluded:

In Radiotherapy Centers of England the Physics Department considers it its duty to go through this thorough and vital study of any new equipment and does not wait until the Clinical Radiotherapist requests that this be done. [174]

On May 16, Mr. Welman wrote Fletcher a letter on G.E. letterhead essentially containing the same material that Grimmett had already put in his May 15 memo to Fletcher and further saying that additional data that Fletcher had requested would be obtained jointly by Grimmett and himself and sent on to Fletcher [175].

Grimmett must have rethought his rather off-hand explanation for the increase in the output of the Maxitron and he set about determining the true cause. On May 16, 1951, he entered into his notebook, "Investigation of increased output of Maxitron 250." And he gave three possible reasons, (1) increase in tube current, (2) shift of target, and (3) increase in voltage. He proposed to investigate the second option by measuring the output at points along the cathode–anode direction and to measure the H.V.L. to distinguish between the other two options. Since the H.V. L. had increased from 2.8- to 3.2-mm Cu he suspected that the increase in output was due to a jump in voltage. The increase in output along the cathode–anode axis for an elongated field, 18 × 3 cm, he found to be 12 % in going from cathode to anode end [140].[1]

Although there was clearly some asymmetry in the beam, it would not have explained the increased output and since the tube current remained constant only option (3) increase in voltage was left.

The next day, May 17, 1951, he had come to that conclusion, "The cause of the overvoltage was traced to the K.V. metering circuit", he wrote in his notebook. He

---

[1] Normally for X-ray machines the output decrease when going from the cathode end to the anode end of the x-ray tube. This is called the 'heel' effect and is more pronounced in diagnostic x-ray tubes than in therapy tubes. It is impossible to know why Grimmett found a rather large opposite effect to the heel effect but possibly indicates adjustments to the tube position in the housing needed to be made.

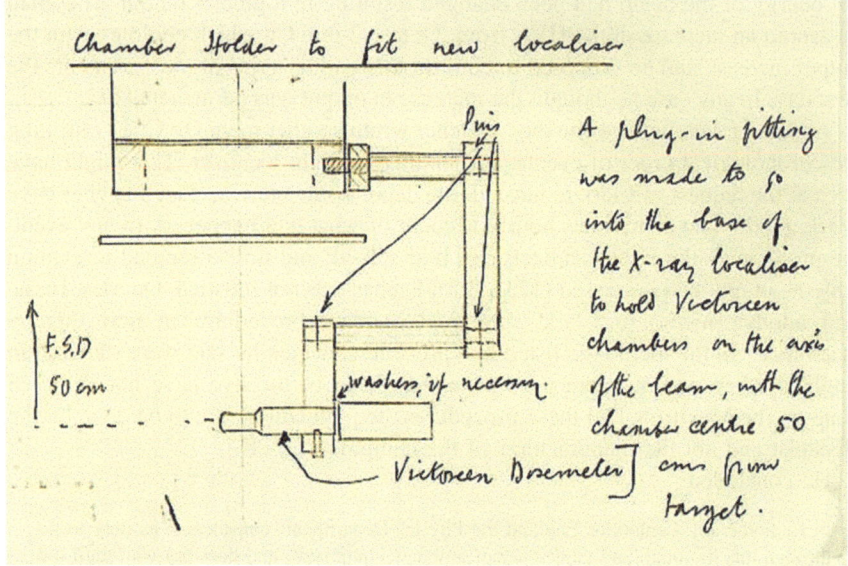

Fig. 9.5 July 25,1950, entry in the notebook about the light localizer

modified the metering circuit so that, "The KV meter was set to read 250 kV when the r-output was ~52 r/min, according to maker figures for an average machine." He installed an A.C. voltmeter across the primary terminals of the transformer as a control to ensure the correct output for a tube current of 30 m.a.

> The voltmeter was left permanently connected as a control over the output. [140]

It was his final entry in the notebook.

Finally, early (7:30 a.m.) on Monday, May 21, Fletcher wrote a memorandum to Clark, apparently in response, once again to a request from Clark for Fletcher to meet in person with Grimmett and resolve the problems with the Maxitron. Fletcher said he was willing to meet with Grimmett but that had not so far been possible since Grimmett had not been available since noon on the previous Thursday. He repeated his accusation that Grimmett had not taken the initiative in getting the Maxitron ready for clinical use and had only done so when pushed to do so by Fletcher and had appeared "completely disinterested and irresponsible."

And: "If Dr. Grimmett is well determined not to assume the responsibilities of a radiophysicist there is no ground for meeting of our minds." This thinly veiled suggestion to Clark that perhaps it was time for Grimmett to be dismissed was reinforced by Fletcher's last paragraph.

> I have had during the last 18 months many heart to heart conversations with Dr. Grimmett when the same fundamental matter repeatedly arose. I have had to reach the conclusion that it has been entirely in vain. I am willing to cooperate and to do team work with Dr. Grimmett but it will be meaningless as long as Dr. Grimmett does not make a true and honest change of heart. [176]

Whether Clark would have agreed to fire Grimmett would never been known. Grimmett died suddenly the following weekend.

The Grimmett' light localizer was eventually made to his design in the hospital machine shop. It was described in detail in a 1953 paper by Grimmett, Fletcher and Moore, published posthumously after Grimmett's death.

> Dissatisfied with the light localizers available commercially, we have developed an improved type, which has several new features, increasing the usefulness and accuracy of the device.

It was a very innovative and unique design, typical of Grimmett and exquisitely made in the machine shop by Bailey Moore. Four features are described, the last was, "…a detachable holder for rapid centering of a condenser chamber for calibration." And a figure showed a Victoreen condenser ion chamber in the holder attached to an orthovoltage machine. The holder, "… brought the effective center of the dosemeter on to the central axis of the beam at a distance of exactly 50 cm. from the target…" [177] (Fig. 9.5).

Chapter 10
Cobalt-60 and the Notebook

Surprisingly, Grimmett's notebook contains no direct reference to the cobalt-60 machine. However, there are one or two projects in the notebook that either directly impacted the design of the unit or involved the use of cobalt-60.

It could be argued that his work on the viewing window glass for the betatron room, mentioned in the last chapter, could be included in this category, but he wrote in November 10, 1950: "Viewing window for the Betatron Room" and did not mention the cobalt room. He was working with the architects at that time on the design of the treatment rooms for the new hospital, but it is more likely that after his sudden death in May 1951, other people made the decision to use the same type of viewing window for the cobalt room as for the betatron room.

Shielding Material

The notebook contains no reference to the cobalt-60 machine. The entry for October 25, 1950, is a table of "Transmission of Cobalt-60 γ-Rays through Different Materials" from experiments of Mayneord and Cipriani, 1947. The entry is in fact a pasted-in table very neatly penned by Grimmett listing a number of elements and materials with their density and mass and linear attenuation coefficients. Three of the listed materials were not from Mayneord and Cipriani's paper. They were tungsten alloy for which Grimmett took the density as 16.5 gm/cm^3, two listings for concrete with different densities which he references to Grimmett and Read without giving any publication information, and Co. The density for the tungsten alloy he had derived the previous day by two methods (Fig. 10.1).

The cobalt-60 unit was designed with a tungsten alloy as the main shielding material, which due to the shortage of tungsten in the United.States. at the time of the Korean War, had to be imported from G.E. in England. But the density of tungsten alloy varies; G.E quoted a range of 16.8–17.0 gm/cm^3. Since this was a quote and he needed to know the exact density of the alloy to be used in the cobalt unit, he had samples of the alloy sent to him in Houston. When they arrived, he

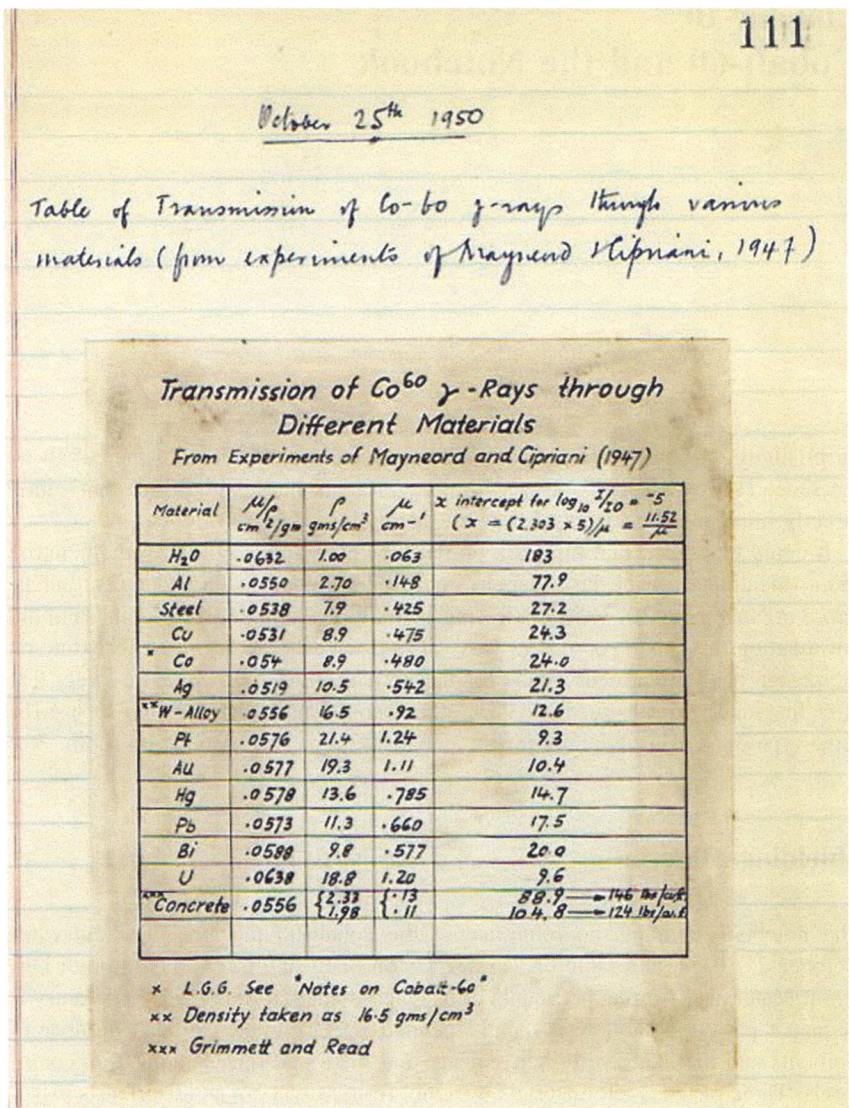

Fig. 10.1 Table on transmission of cobalt-60 γ

undertook, on October 24, 1950, three separate methods to determine the density of the samples.

1) Archimedes principle: A 500.5-g sample was weighed in air and weighed in water, and the difference gave the weight of the water displaced. The temperature of the water was measured, and therefore, its density was known, and the volume of the water displaced was calculated, and since this had to be the volume of the tungsten, the density was easily calculated.

[A simple error was made in this calculation, in calculating the volume of water displaced a decimal point was shifted, and the volume was recorded as 3.04 cc. In fact, it was 30.4 cc. However, he used the correct value in the next line to determine the density of the tungsten alloy so the answer came out correctly. (This was a very rare occurrence for Grimmett, usually he was very careful in what he recorded.)].

2) Volume displaced in a measuring cylinder for a 500-g sample, which was found to be 29.7 ml.
3) When calculating the volume for the 500-g sample by measuring its dimensions, Grimmett noted this was not accurate and it yielded a value of 26.9 ml, 10 % different from the above value, and Grimmett did not use it.

The first two methods yielded values of 16.43 gm/cm^3 ± 0.1, 16.7 gm/cm^3 ± 0.2, and he wrote: "Probable mean value ≈ 16.5 gm/cm^3," and this was the value he wrote in the table.

The notation for Co says, "L.G.G. See 'Notes on Cobalt-60'."

This would indicate that Grimmett had another notebook in which he kept information on cobalt-60 and perhaps his thoughts on the design of the unit. But this notebook has never been found.

Experiments with Cobalt-60

In 1950, Grimmett received a small cobalt-60 source.

On November 28, 1950, he wrote in the notebook: "Preliminary estimate of strength of nominal 2 Ci Cobalt-60 slug by means of Victoreen dosemeter." For the next three days, he made measurements using the experimental arrangement shown in Fig. 10.2. He made measurements with Lucite filters of various thicknesses over the hole in the container as a measurement of the dose buildup, and he used a thin aluminum filter to stop as he wrote, "Soft γ's and electrons from Pb wall." He then extrapolated the data back to zero thickness to take account of the attenuation in the filters. He measured 0.86 r/min at 19.9 cm. He wrote down what must have been the accepted value of the exposure rate constant for Cobalt-60 at that time as:

"1 mc cobalt-60 gives 13.5 r/hr at 1 cm" (The present value is 13.07.) and derived a value of 1510 mCi for the source. This was then corrected for absorption taking account of the attenuation in the aluminum filter and self-absorption in the cobalt source, using the linear attenuation coefficients from the table in the notebook. He calculated the source strength as 2.01 ± .04 Ci on November 30, 1950. At some later date, he made a small correction to the calculations. The thickness of the aluminum filter had been left out of the determination of the distance to the center of the chamber. It was a small correction, and the final determination was 2.05 ± .04 Ci.

He was clearly thinking about how he would go about calibrating the cobalt unit when it was completed.

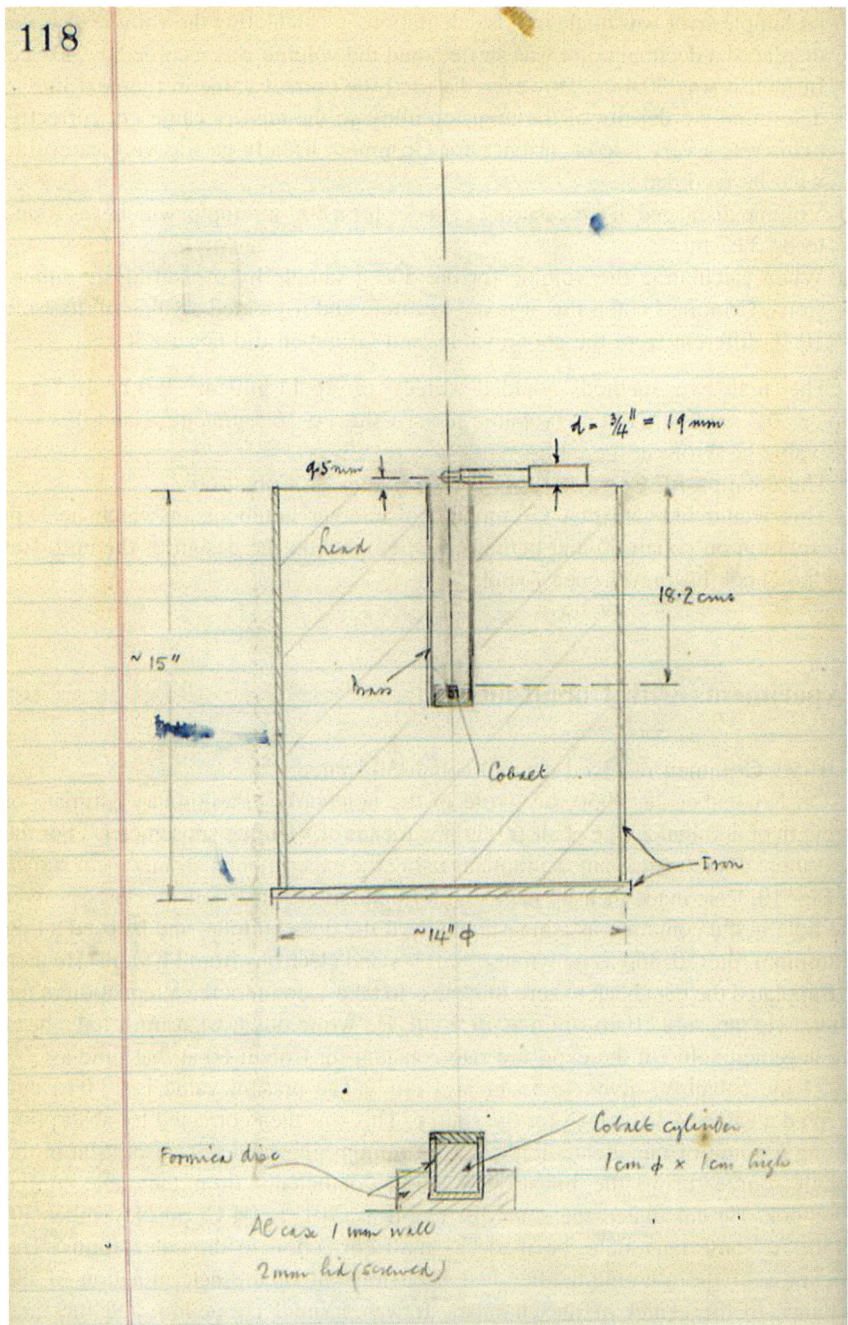

Fig. 10.2 The experimental setup to measure the source strength and the dimensions of the source

Experiments with Cobalt-60

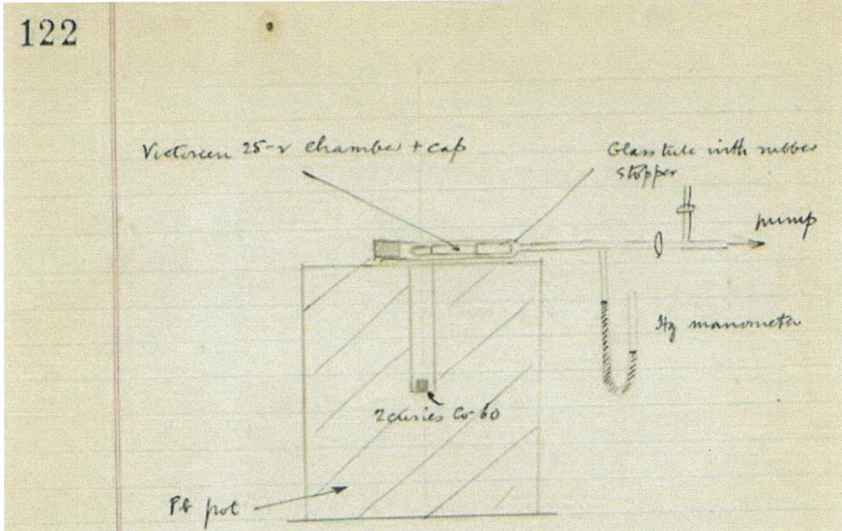

Fig. 10.3 The experimental setup to measure the chamber response versus air pressure

There was some concern that the Victoreen chamber used did not follow variations of atmospheric pressure. The next day, December 1, 1950, Grimmett modified the previous experimental setup to test the chamber. He placed the chamber in a glass tube, sealed at one end with a glass stopper and the other end was connected to a mercury manometer, to record the pressure and to a pump to vary the pressure (Fig. 10.3).

The response of the chamber, in terms of r/min, was plotted against the pressure in terms of millimeters of mercury (mm Hg). The atmospheric pressure at the time the cobalt source was calibrated had been 736 mm, but for this experiment, the pressure was only varied from 682 mm to 716 mm. Grimmett concluded that:

>...incomplete, no data for atmospheric pressure, and pressure range too small.
>Data do not differ significantly from the theoretical expectation. Must repeat.

On the graph, he noted that: "All the points fit curve within ± 1 part in 75 except 1" (Fig. 10.4).

Contamination

No other experiments with the 2 Ci source are recorded, and no other reference to cobalt-60 is made in the notebook. However, tucked into the back pages of the notebook were three pieces of paper, two graphs on semi-log paper and part of a filing folder that had been torn in two containing some data.

The note on the torn filing folder says, "Radiation from cobalt 'pot'," and readings taken with a Beckman radiation meter at various positions on the pot. The

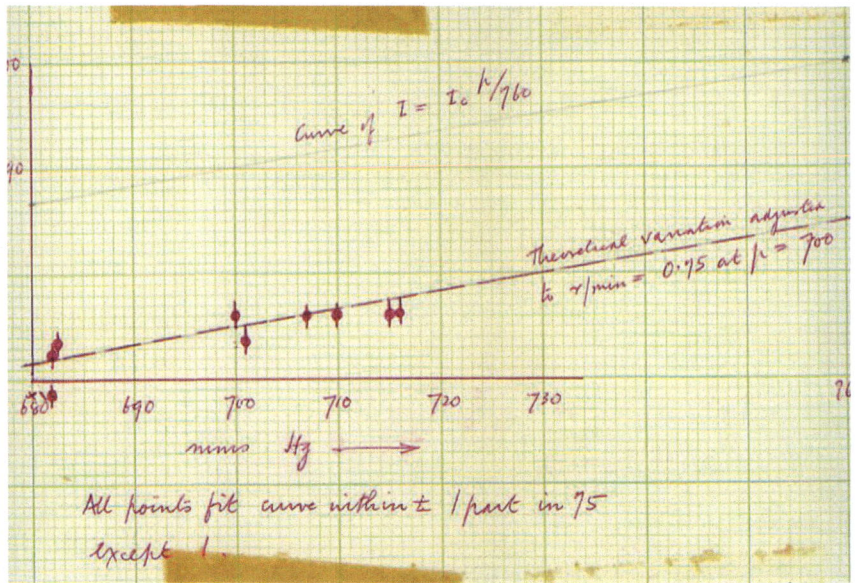

Fig. 10.4 Results of the experiment to measure chamber response versus pressure

readings vary from 8–18 div. on the 20 scale. But there is no way of knowing what that means. However, the two graphs give some further insight. One is marked, in Grimmett's handwriting, "Cobalt-60 Contamination." This graph, on semi-log paper is a plot of filter thickness in mg/cm^2 versus counts/second. It is a two component curve. The first part shows radiation completely stopped by 60 mg/cm^2 and probably represents low-energy electrons being emitted by the contamination. The other component but with data only out to 120 mg/cm^2 shows no attenuation at all and may represent a high-energy γ-ray component.

The other curve is just as intriguing. Again the notation on, it is in Grimmett's handwriting and says:

> 'Cobalt' contamination from HCl solution (filtered) of active material from forceps used to handle 2 curies of Co-60
> Sept 25 1950.

Clearly, the forceps used to handle the small cobalt source had become contaminated, and the contamination had been dissolved off with hydrochloric acid. This graph is of time versus counts and is again a two component curve. There is a short half-life component that is noted on the graph paper as 2.04 h and a longer component with no decrease out to seven hours.

It is clear, therefore, that Grimmett received the cobalt source at least a couple of months early than was indicated by the entries in the notebook, and at this early stage of distribution of radioactive sources from Oak Ridge and elsewhere, there were problems with contamination and perhaps purity of the sources.

Chapter 11
Cobalt-60 in Perspective

Like most successful innovations, there have always been numerous claims to be the first to have made the suggestion to use cobalt-60 for teletherapy. At a symposium on teletherapy in 1956 at ORINS, Brucer said:

> Depending upon whether one is a Canadian, an American, Russian or even an Englishman, Co^{60} production started (for teletherapy purposes) in one of these countries. I have selected 1951 as the first year in which Co^{60} was produced for teletherapy; however, as time goes on, the date for the first Co^{60} machine goes back further and further in history…I have in my files of newspaper clippings absolute proof of at least fifty "first" Co^{60} machines. All these people are liars, of course, because we have the first machine. [178]

But the situation was quite complicated because the claims for priority can be broken down into subsections. Who was the first to suggest replacing radium with an artificial radioactive isotope? Who was the first to suggest that cobalt-60 would be the ideal isotope to replace radium? Who was the first to suggest that sufficient amounts of cobalt-60, for such use, could be prepared in a nuclear reactor? Who was the first to design a treatment unit to specifically use cobalt-60? Who was the first to build such a unit? Who was the first to send stable cobalt-59 to a reactor for activation for use as a source in a treatment unit? Who treated the first patient on a cobalt-60 treatment unit? And finally, who developed the clinical use of the cobalt-60 machines making them one of the most effective tools in the treatment of cancer in the latter half of the twentieth century? There are, therefore, many sub-areas for individuals to claim to have been the first and many did and sorting out the priorities 60 years later is not easy. In the scientific world, such claims rest upon publications in recognized scientific journals to establish priorities. Since much of the cobalt-60 story evolved out of the Manhattan project during World War II and publication in the open literature was not often possible, establishing priority is quite difficult.

Because of this, there developed the myth that the Canadians and the Americans (specifically the MDAH/ORINS group) were both claiming to have invented the cobalt unit. But this was never the case.

The clinical use of the MDAH/ORINS unit ultimately depended upon the building of the new hospital in Houston. The temporary location for the MDAH on the Baker estate had no suitable building to house the unit and the cost of constructing one, which in any case would only have been used for one or two years, would have been too much, especially since the funds for the new hospital building had not yet been fully raised. This was clearly recognized from the beginning. In the July 1950, contract between MDAH and ORINS, ORINS was to construct a suitable building where the machine could be tested, the physical characteristics of the beam determined, and radiobiological studies undertaken. If these studies were successful and a determination made that the unit should be released for use on human beings, a request would then be made to the AEC to do so. If the commission approved then the unit would be transferred to MDAH in Houston for the commencement of patient treatments. All of this would take time, and one of the biggest delays was due to the problems associated in obtaining the 1,000 Ci source. When Grimmett and Brucer met with Aebersold in September 1949 at the Oak Ridge National Laboratory, they believed that ORNL could supply them with the 1,000 Ci source. However, when Dr. Lough reported on the availability of cobalt-60 sources from the ORNL reactor at the February 1950 meeting in Washington D C on Cobalt-60, it was immediately clear that ORNL could not supply the sources. Grimmett was so concerned about this that he addressed a special section in his report on the meeting to this subject and concluded that the Chalk River reactor in Canada was the only place were the sources could be activated, and he would "make do in the mean time with the most powerful source having the same dimensions which Oak Ridge can supply" [113]. By May 1950, the unactivated sources were sent to Canada, and in June 1950, they were inserted into the high-flux nuclear reactor at Chalk River along with two sources for the Canadian machines.

It was at the February 1950 meeting that the MDAH/ORINS group found out that two groups in Canada were completing construction on cobalt units. One designed by Harold Johns of the Saskatoon Cancer Center and manufactured by the Acme Machine Shop in Saskatoon, and the other designed by R.F. Errington of the Eldorado mining and Refining Company where a unit was being built for the Victoria Hospital in London, Ontario.

The story of these units has been well told in a series of articles in the publication InterACTIONS, the Canadian Medical Physics Newsletter [179–181].

By the time the MDAH/ORINS sources arrived in Canada, both these projects were well underway. All three machines were designed for 1,000 Ci. The three sources were loaded into the reactor in the summer of 1950, and it was anticipated that they would be up to full strength in 10 months. The Saskatoon source was delivered in July 1951 and installed into the unit on August 17, 1951. The London, Ontario source was delivered in October 1951 and installed on October 23, 1951; they treated their first patient on October 27, 1951. The Saskatoon group took longer to test their machine before patient treatment, treating their first patient on November 8, 1951. The race to treat the first patient on a cobalt-60 unit was between the two Canadian groups, with Errington understanding the commercial importance to his company to be first [182]. As recounted in Chap. 8, the MDAH/ORINS source was put back into the reactor for an additional six months. In the mean time, the unit was shipped from Milwaukee to Oak Ridge and was loaded with the 200 Ci source that had been irradiated at Oak Ridge for Dr. Max Cutler of the Chicago Tumor Institute. The Chalk River source was not shipped to Oak Ridge until July 1952, and the unit underwent further testing before being shipped to Houston in September 1953 when the construction of the new hospital was far enough along to allow it to be installed. The first patient was treatment on February 22, 1954.

However, this was not the first patient treated by cobalt-60 in the U.S.A. That recognition goes to the Los Angeles Tumor Institute.

In a September 1953 paper in *Radiology*, Russell Hunter Neil (the physicist at the Los Angeles Tumor Institute), William Costolow and Orville Meland (the radiologists involved), described their unit and the treatment of the first patient with it. They had been using a 4 g radium teletherapy unit, and some time after 1948, they began to give consideration to the design and construction of a cobalt-60 unit. Neil, who attended the February 1950 meeting in Washington D.C. on cobalt units, heard that Oak Ridge was activating a large number of smaller sources, and he decided to use a combination of these to make a 1,000 Ci source. It consisted of six stacks of 18 pieces each for a total of 108 smaller sources forming a cylinder 4.33 cm tall and 3.5 cm in diameter; each individual cobalt source was sealed in a stainless steel tube with brass ends. On February 25, 1952, the total activity was 1,080 Ci of cobalt-60, weighing 181.74 g for a specific activity of 5.94 Ci/g. This yielded an output of 32 r per minute at a treatment distance of 70 cm. With such a big source, the penumbra was very large, and there were shielding problems. The first patient was treated on April 23, 1952 [183].

A reasonable chronology for the development and initial use of cobalt-60 units can, however, be derived from all the information available. It should be noted here what is understood by the term "cobalt-60 unit."

Although the initial idea might have been to replace the radium in radium teletherapy units with an equivalent activity of a suitable radioactive isotope, such an arrangement would have ended up with all the disadvantages of such units; poor beam penetration due to the predominance of the inverse square law at short treatment distances, so negating the benefit of the megavoltage γ-rays, large source size resulting in a big penumbra, low activity resulting in a low dose rate and extended treatment times. With the larger amounts of activity that were available with cobalt-60 (1,000 Ci), the cobalt units could be designed and used for larger treatment distances. Although Grimmett's design was for 70 cm, it eventually was used at 75 cm and for the commercial units 80–100 cm became common. Even at these distances, the output was good being comparable to the 250 kV X-ray units with which most of the world's radiotherapy was being done at the time. Even though the source size (cm) was always bigger in a cobalt unit compared to the small focal spot (mm) of an X-ray machine, the treatment distance and careful design of the collimation greatly reduced the effect of the penumbra. From the beginning, therefore, the cobalt units were considered to be a megavoltage replacement for the kilovoltage X-ray machines not as a replacement for the teletherapy radium units.

Only the chronology of the development of such machines is given. The other approach was tried but without much success. Bryant Simons & Co of London, who built the commercial version of Grimmett's pneumatic transfer radium unit, offered a unit with 10–150 Ci of cobalt-60 as the source [184]. This in fact was the first approach that Grimmett and Fletcher proposed for M.D. Anderson Hospital. They included a line item in the budget they submitted in mid1949 to purchase the equipment from Bryant Simons in London, but without a source, leaving open the option to get either a radium or cobalt-60 source. With the rapid development of the other approaches to cobalt-60, the idea was quickly dropped. In a footnote in Chap. 8, reference was made to the iridium-192 unit designed by Freundlich (who had been a member of the British group of scientist in Canada during the war) which was a short treatment distance, low output, kilovoltage γ-ray machine but which in fact did treat the first patient on equipment using an artificial radioactive source. Cs-137 treatment machines were also built and have had some commercial success. The half-life of Cs137 is 30 years, which made such equipment attractive since the source did not have to be changed as often. But the γ-rays from cesium-137 only have energy of 660 keV, and such machines are generally used at short treatment distances for head and neck tumors, where beam penetration is not so important. The chronology includes none of these developments.

Nor are the activities of Lewis Strauss included. Strauss was a wealthy financier with a taste for physics and politics. In an article by T. A. Heppenheimer titled "How to detect an Atomic Bomb" in Invention and Technology in the spring of 2006, Heppenheimer wrote of Strauss:

> His involvement with physics dated from the mid-1930s, when both his parents died of cancer. 'I became aware', he later wrote, 'of the inadequate supply of radium for the treatment of cancer in American hospitals'. He searched for ways to produce a radioactive isotope of cobalt as a substitute, and this led him to Leo Szilard. As the two men grew close, Szilard made Strauss, who had never gone to college, one of the first nonscientists to learn of the prospect of an atomic bomb [185].

On the Lewis L. Strauss' page on the internet [186], a few more details are given. He had heard from his physicist friends, at the time of his parent's death, about the possibility of bombarding cobalt with high-energy subatomic particles to create radioactive cobalt-60. Strauss, it was said, "saw this as a way of making cobalt-60 cheaply and providing it to hospitals as a replacement for radium," and he looked for ways to build the necessary accelerator and distribute the isotopes as a memorial to his parents. But the project never went very far, and when World War II started, it was dropped, and the advent of the nuclear reactor made the project obsolete. In 1946, Strauss was appointed by President Truman as one of the commissioners of the new Atomic Energy Commission. It is hard to see how Strauss could have made the connection between the announcements of radioactive cobalt-60 in the 1930s and using it to replace radium. The amount of cobalt-60 produced at that time by various means was minute and was detected only by its radioactive decay. There were also considerable questions at as to its half-life, and the energy of the γ-rays as discussed in Chap. 6. However, he may have seen a Letter to the Editor of Physical Review by Livingood and Seaborg on the long-lived radio-cobalt isotopes [187]. They had been following the decay of chemically separated samples of radioactive cobalt and had the energy about right 1.3 MeV, but believed there was a long-lived isotope of cobalt with a half-life of 10 years. They admitted, however, that there could be competing radioactive isotopes from traces of iron or nickel in their cobalt oxide samples. They also discussed the production of radioactive cobalt in accelerators, but again stressed there were many competing interactions taking place. If Strauss had been aware of the Eve and Grimmett paper in Nature [53] where Grimmett had suggested replacing the radium in teleradium units with an artificial radioactive isotope, saying that radio-sodium might be a possibility, Strauss might have thought that with Livingood and Seaborg reporting a megavoltage γ-ray and a half-life that could be 10 years for cobalt-60, it would be an excellent candidate. It was, however, a far stretch to think that suitable amounts of the isotope could be produced with an accelerator. This, of course, is speculation, but when the question of approving the use of reactor-produced radioactive cobalt-60 as a source for teletherapy machines for cancer treatments came up for discussion by the AEC Commissioners. Strauss must have felt some justification for his earlier endeavor.

Chronology

1937 January: Nature, Grimmett suggested an artificial isotope might be found that could replace radium in teletherapy machines. He also intimated that if enough activity of the isotope became available, then extended treatment distances could be used [53].

1941–1942: Livingood and Seaborg's (1941) [80] and Nelson's papers (1942) [81] on cobalt-60 were published.

1944–1945: Grimmett reading the physics journals of the time saw papers on cobalt-60 and realized it might be the radioactive isotope to replace radium [76]. Who else read these papers and came to the same conclusion is not known, although Fletcher for one made that claim [188].

1946: Mayneord who had replaced Mitchell as adviser on biological and medical research at Chalk River in 1945 gave a series of lectures on the physics of radiotherapy at the Toronto General Hospital and spoke enthusiastically about cobalt-60 as a source for radiotherapy machines. Johns was in the audience.

1946 December: J.S. Mitchell's paper in the British Journal of Radiology recognizing cobalt-60 as a viable replacement for radium and the practicality of its production in a nuclear reactor was published in December [82].

1947 October: shipment of a small Cobalt-60 source to the University of Saskatchewn [179].

1947 November: Paper published by Mayneord and Cipriani in the Canadian Journal of Research on the absorption of γ-rays from cobalt-60 [13].

1948–1949: Design and construction began on a Cobalt-60 applicator for use in cancer therapy by Russell Neil at the Los Angeles Tumor Institute [183].

1949: Three Cobalt-60 units designed independently: H.E. Johns in Saskatoon, R.F. Errington at the Eldorado Mining and Refining Company for the Victoria Hospital in London, Ontario and L.G. Grimmett for the MDAH/ORINS.

1949 August: H.E. Johns and T.A. Watson submit written request for a kilocurie source to be produced by the Chalk River reactor [180].

1950 May: Grimmett presents a paper and an exhibit on the "1,000 Ci Cobalt-60 Irradiator" at the MDAH fourth annual Cancer Research Symposium [118, 119].

1950 June: Eldorado Mining and Refining Company and MDAH/ORINS requested kilocurie sources from Chalk River. Three sources were loaded into the NRX reactor at Chalk River. Anticipated time to reach a 1,000 Ci was 10 months.

1950 July: Fletcher presented a paper in Paris at the Fifth International Cancer Congress detailing the design and operation of the MDAH/ORINS unit [121] and at the Sixth International Congress of Radiology in London. At that meeting, Errington showed a model of the Eldorado Mining and Refining Company's cobalt unit.

1950 November: The 2 curie source at MDAH is used for experimental purposes [140].

1950: Grimmet publishes his paper on "A 1,000 Ci Cobalt-60 Irradiator" [120] in the October–December issue of the Texas Reports on Biology and Medicine.

1951 July: Source delivered to Saskatoon, installed August 17 [179].

1951 October: source delivered to London, Ontario, installed October 23 [180].

1951 October 27: First patient treated London, Ontario [181].

1951 November 8: First patient treated Saskatoon [180].

1951: Johns et al. paper on 1,000 Ci cobalt-60 machine in Nature [189].

1952 April 23: First patient treated at the Los Angeles Tumor Institute.

1952 July: MDAH/ORINS source delivered to ORINS.

1953 September: the new M. D. Anderson Hospital in Houston was far enough along in construction to allow the cobalt-60 unit to be shipped from ORINS.

1954 February 22: First patient treated at MDAH.

Although this chronology stops at the treatments of the first patients on the initial cobalt-60 units, that is far from the end of the story. For cobalt-60 teletherapy to be successful, two other criteria needed to be fulfilled. The treatment themselves had to show improvement over other treatments prevalent at the time, and the cobalt machines had to prove economically viable.

The latter development proceeded rapidly, long before the clinical results were in. Although a few additional units were built by individual institutions for their own use such as the unit at the University of Louisville for Herbert Kerman who had been on leave at Oak Ridge as the physician with the MDAH/ORINS project [190], commercial companies immediately entered the field. At the Atoms for Peace conference in Geneva in 1955, Brucer reported that there were approximately 120 cobalt units in use around the world, four years later the number was estimated to be 300–350, with 60 being used in Japan and at least 16 in the Soviet Union. Canada had produced 166 Cobalt-60 sources and the United States 171 sources [191].

In 1959, the International Atomic Energy Agency (IAEA) produced a global Directory of Teletherapy Equipment. It listed 46 models built by 18 firms in nine countries: Canada, France, Germany, Italy, Japan, the Netherlands, Sweden, the UK and the USA. In addition, units had also been built in the Soviet Union, several Eastern European countries and in China [191].

This is a remarkable expansion in 10 years in the number of manufactures, countries using cobalt-60 units, and countries supplying cobalt-60 sources, since the initial February 1949 conference on cobalt-60 teletherapy in Washington D.C., when there were no finished units, no patients had been treated on a cobalt-60 unit, and there were questions about who could supply the radioactive cobalt-60. By 1965, cobalt-60 sources were being supplied by at least five different countries, U.S.A., Canada, the UK, Australia and the Soviet Union [184].

In 1976, the IAEA reported that there were 2365 units installed worldwide, but by 2012, the number was down to 1625. The reasons for this decline will be discussed later. There was a wide range in the types of cobalt-60 units built. Some were mounted on a stand others as rotational units where the whole treatment head could be rotated $360°$ around the patient. Source strengths varied greatly from the Deka-Curie unit manufactured by Elma-Schönander of Stockholm to the 10,000 Ci source from Shimadzu in Japan and everything in between. Treatment distances were also varied. For the units on a stand, the treatment distance could vary, usually in the range of 50–80 cm. For the rotational units, the center of rotation, called the isocenter, was usually 80 cm from the source although modern units can have isocenters at 100 cm.

Although G.E. had built the unit for the MDAH/ORINS group, they decided not to enter the commercial market for such machines. As noted in Chap. 8 that machine treated patients until 1963 when it was decommissioned and eventually donated to the Smithsonian Institute in Washington D.C., Harold Johns sold the rights to his machine to the American company, Picker X-ray in Cleveland, Ohio. In a reply to an inquiry from E.R.N. Grigg about Picker's share of the world market, Henry Picker replied in a letter dated December 4, 1963, that:

> It is very difficult for us to be certain about the number of cobalt units we produce compared to other companies…However, we can be reasonably sure that we produce over one-third of all Cobalt units produced in the world [192].

John's original machine was used for 21 years and was decommissioned in 1972. The machine at the Los Angeles Tumor Institute was constructed from components that had been readily available. Neil, who designed the unit, did not stay in medical physics, and although it was used on a daily basis until early 1962, the unit had deteriorated, and the source decayed so that its output was too low to use. Since the source had been constructed from multiple smaller cobalt-60 sources, the unit was not compatible with the newer source configurations, and the machine was decommissioned.

The unit built by Eldorado Mining and Refining Ltd. for Victoria Hospital in London Ontario was a commercial undertaking from the beginning. In 1949, the company had obtained the rights to distribute cobalt-60 from Chalk River to commercial and industrial users, although Chalk River retained the right to distribute cobalt-60 and other isotopes to research institutions and universities. Errington was head of the Commercial Products Department at Eldorado Mining and Refining and saw the potential for cobalt-60 in the medical field. On his staff was Don Green a graduate in engineering physics. Errington had read Mayneord and Cipriani's paper on cobalt-60 and like others realized that a cobalt-60 treatment unit for cancer therapy was a possibility. He and Green discussed this with Ivan Smith, head of the cancer clinic at Victoria Hospital in 1949 and were encouraged to pursue the idea, and by late 1949, they had received company approval and funds to go ahead. Eldorado Mining and Refining Ltd. were now in a new business.

The construction of the first unit was contracted out to Canadian Vickers in Montreal, but Errington was not particularly happy with the arrangement, and he made plans to manufacture his own units. Errington knew the value of publicity and shipped a model to England for display at the International Congress of Radiology in London where he demonstrated the machine, called the Eldorado A, to Queen Elizabeth (the wife of King George IV). It was at this congress that Fletcher gave his presentation on the MDAH/ORINS cobalt-60 unit. By the end of 1952, they had received orders for four units, and the company was working on a rotational unit—the Theratron B. (Vertical stand units were called the Eldorados and the rotational units Theratrons.) The first rotational unit was installed in the Francis Delafield Hospital in New York in May 1953.

The Chalk River facilities were part of the National Research Council (NRC) of Canada operations but as Chalk River expanded a new arrangement was necessary and on April 1, 1952, a new crown corporation, Atomic Energy of Canada Limited (AECL), came into being to manage Canada's atomic research. Five months later, AECL took over Eldorado's Commercial Products Division. AECL became one of the most successful cobalt unit companies in the world. When the Johns machine in Saskatoon was decommissioned in 1972, it was replaced with an AECL unit, and at the M.D. Anderson Hospital in Houston, Fletcher eventually had five AECL units, one Eldorado and four Theratrons.

In 1988, a move was made to privatize the company, and the medical products division became Theratronics International Limited, and today, it is one of the Best companies known as Best Theratronics. It still sells cobalt-60 units. The company literature says that over the years, it has installed more than 1,500 Theratrons and supplies over 80 % of the world's γ-based external beam therapy systems.

An extensive account of the Eldorado Mining and Refining Ltd cobalt-60 story can be found in Paul Litt's book, Isotopes and Innovation [191].

This commercial success would not have continued without the fact that cobalt-60 treatments proved to be one of the most successful and practical anti-cancer therapies of its time. From the beginning, it was hoped that this would be the case due to the characteristics of the cobalt γ-ray beam, and this resulted in the early enthusiasm.

Grimmett had discussed the advantages expected from the use of cobalt-60 in his 1950 paper[120]. First, the energies of the γ-rays of 1.33 and 1.17 meV were higher than the X-ray energies from the kilovoltage X-ray units then in use. This allowed the energy from the radiation to reach further into the body resulting in deeper tumors being treated. Second, the dose to the skin is lower than for the X-rays, a phenomenon called skin sparring. For X-rays, the maximum dose (100 %) is at the surface, and for cobalt-60, the maximum dose occurs a depth of 0.5 cm, and the surface dose is in the range of 30–50 %. For kilovoltage X-ray therapy, the amount of radiation that can be given is limited by the skin reaction, which is like severe sunburn. Although there is skin reaction with cobalt-60 it is much less and higher doses can be given and the higher the dose the greater the possibility of eradicating the tumor and with the greater depth dose higher doses can be delivered to greater depths. Third, at these energies, the radiation is more uniformly absorbed by all tissues resulting in an even distribution of the dose over the tumor. At kilovoltage energies, bone preferentially absorbs more energy than muscle and other tissues. When too much radiation is absorbed by the bone, bone necrosis can occur, and if the tumor lies behind bone, the tumor will receive less radiation than required and the tumor might recur. These problems were minimized by the radiation from cobalt-60. All of this contributed to the expected advantage for the cobalt units, and this proved to be the case.

There was another, less tangible, but nevertheless just as real an advantage for the cobalt units, the beam from machine to machine was the same allowing meaningful clinical trials to be done, and in the forefront of carrying them out was Gilbert Fletcher at the M.D. Anderson Hospital in Houston. Fletcher was one of a new generation of radiotherapists who entered the field after World War II.

He was 36 years old when Clark appointed him a traveling Anderson fellow, and he was in a hurry to get establish as a radiotherapist. He believed there was a great opportunity to do so especially with the treatment of head and neck cancer, gynecologic cancer and breast cancer.

For clinical trials to succeed, they had to be very carefully controlled, which was difficult to do with kilovoltage X-rays. There was no standard X-ray machine and no standard X-ray beam. The quality (energy) of the beam was dependent upon the applied voltage to the tube, the filtration inherent in the particular design of that X-ray machine and the filtration that was added to help determine the energy of the X-ray beam. In addition to those factors, the output was also dependent upon the current through the tube. Fletcher demanded a very stable output from his X-ray units, which was difficult to obtain. This was especially true at MDAH's temporary quarters on the Baker Estate where the voltage supply was not particularly constant. Grimmett did his best to meet Fletcher's demands but could never convince Fletcher that he was doing so and, as described Chap 9, this led eventually to the breakup of their relationship.

But with cobalt-60, the beam was the same from each machine, and once the output was calibrated for a given machine, the output could be determined thereafter by allowing for the decay of the radioactive cobalt 60. With a half-life of 5.26 years, the output of every cobalt unit decreases by 1.1 % per month. Fletcher was a very demanding person with a genius for organization and technique, and with a large patient population, he used the advantages that cobalt-60 offered to amass a wealth of clinical data. He expected a strict adherence to treatment technique, reliance on accurate dosimetry, constant review during treatment and extended and thorough follow-up on each patient after treatment. Finally, the statistical analysis of the results had to be faultless, and with an advanced degree in mathematics, he made sure that it was. He published extensively and wrote a textbook on radiotherapy [193], so that the clinical advantages and improved results with cobalt-60 were quickly known.

However, the use of cobalt-60 machines was not without its problems.

Several thousand curies of radioactive material is always potentially dangerous, and the source is never "off", it can only be heavily shielded to make it safe to be around. There were several times with the Grimmett unit when the source failed to return completely to the off position at the end of treatment, and a manual override had to be used when this happened. For the later designs of the Eldorado and Theratron, the source was in a draw that moved horizontally. A red rod, attached to the source draw would protrude out of the treatment head indicating that the beam was on. If at the end of the treatment the source did not return to its off position, the rod would still be visible. The procedure was then to get the patient off the table and take a rod, provided with the machine that was hollow at one end and had a tee bar at the other. The hollowed out portion was placed over the protruding red rod, and the operator pushed on the other end with the tee bar until the source was returned manually to the off position. On these machines, the source draw was moved by compressed air. Had Grimmett lived he would have found great satisfaction that the cobalt units relied on compressed air to turn the units on, just like his pneumatically controlled radium unit.

On one occasions, the source on the Grimmett unit leaked radioactive cobalt oxide, which is a fine white powder, onto the treatment couch and floor of the treatment room, which was then carried on people's shoes around the hospital. Although the amount of radiation involved was small, everything had to be decontaminated. This led to an improved source container design [194].

The sources now consist of very small nickel-plated cobalt-60 pellets, a few millimeters in length and diameter. These are then poured into a double-encapsulated cylindrical container made of low carbon stainless steel. A special assembly procedure is used to ensure uniform source density, and stainless steel spacers are placed above the pellets to keep them secure. The cylinder is then closed with a stainless steel lid, which is welded in place. The sources come in two sizes, 1.5 and 2.00 cm diameter. In some of the earlier sources, the packing was not tight enough, and the pellets moved when the machine was rotated effecting both the output and dose distribution. There are sources, however, which use solid cobalt-60 rather than pellets.

The above incidences were infrequent and although they had to be corrected generally did not result in serious problems. However, the potential for serious problems is inherent with cobalt-60 units.

If the decay of the source is not accounted for correctly, disastrous results can occur, as illustrated by the tragic events at Riverside Hospital in Columbus, Ohio 1974–1976 [195]. The unit was initially calibrated correctly, and the decay of the source was used to determine the dose rate and treatment times thereafter. Rather than calculating the decay, the output was plotted on semi-log graph paper, with the log y-axis for the dose rate and the linear x-axis for time. On such graph paper, the output is a straight line. When the line reached the edge of the graph paper, the plot was continued on another sheet of graph paper, but this time incorrectly on linear graph paper, although the extrapolation was continued as a straight line. Since the linear y-axis did not correspond to the log y-axis, the straight line extrapolation resulted in increasing incorrect output with time. The error resulted in the dose rate being underestimated by 10–45 % resulting in a corresponding overdose to the patients, which increased almost linearly with time. This occurred over a 22-month time period during which the output of the cobalt unit was not measured. 426 patients received significant overdoses. During a followup period of 1–3 years after treatment, a high proportion of the patients developed significant, often life-threatening complications. As a result of this incident, the Nuclear Regulatory Commission issued extensive regulations on the training requirements, and quality assurance procedures required when using cobal-60 machines.

Problems can also occur with the disposal of the radioactive source. Even after the source in a cobalt-60 unit has decayed to where it is not economical to use for patient treatment, it is still highly radioactive and proper disposal of the source is required. Normally, the company replacing an old source with a new one takes the old source. However, sometimes, this is not the case. In 1977, a cobalt-60 teletherapy unit was illegally imported into Ciudad Juárez, Mexico, just across the border from El Paso, Texas. The unit was never reported to the Mexican authorities and was stored in a warehouse for 6 years. In December 1983, maintenance staff at the warehouse having no idea what it was or the potential danger associated with it became interested in the unit for its scrap metal value. It was loaded onto a truck, where the source container was perforated, and was driven to a junkyard where it was sold as "valuable" metal pieces. The source strength was about 430 Ci consisting of about 6,000 cobalt-60 pellets, which as mentioned above are about 1 mm in size. These pellets were then scattered around the junkyard, mixed with other metal scrap and eventually on other trucks moving scrap out of the junkyard. The main purchaser of the scrap used it to build reinforcing rods for construction, another purchaser made metal table bases from it. Eventually a truckload of tables passing the Los Alamos Nuclear laboratory in New Mexico triggered their radiation monitors. The authorities were notified, and the origin of the contamination was determined. The contamination was extensive, 4,000 people were exposed, 814 buildings were found to have levels of radiation above the regulatory limit, and eventually 37,000 tons of rods, metallic bases, scrap iron, barrels with pellets, contaminated material, earth and material in various stages of manufacturing process had to be stored and allowed to decay [196].

In later years, the whole question of accountability of the radioactive sources and making sure that they do not fall into wrong hands as part of anti-terrorist concerns has led to more restrictions on the use of cobalt units.

But the very success of the clinical use of the cobalt units sowed the seeds for the decline in its use, and the above kinds of problems did not help. The cobalt-60 teletherapy unit was not the only development that leads to new kinds of treatment machines after World War II. There were several others. It will be recalled that just before the war, Grimmett started cooperative work with Imperial College in London to build an electrostatic generator called a Van de Graaff accelerator, which, due to the war and Grimmett's dispute with the Medical Research Council (MRC), he did not get to finish. At the January 2, 1948, meeting of the Hospital Physicists' Association in London, which Grimmett attended, Lamerton gave a report of the status of radiotherapy and medical physics in the United States, and Grimmett made notes in his diary. He made a note of the fact that there was an electrostatic generator (e/s) at MIT for radiotherapy purposes. He wrote in his diary:

MIT
6ftx3ft 2 MeV e/s generator
servicing risk? 50,000$ [197].

On his way to Houston, therefore, he made the extra effort to extend his journey by a couple of days so that he could go to Boston to talk with John Trump, at the High-Voltage Research Laboratory at MIT, who was also the technical director of the High-Voltage Engineering Corporation which built Van de Graff accelerators. Trump was a proponent of using these accelerators for radiotherapy.

It will also be recalled that on his trip to the United States in 1937, he had visited with Kerst at the University of Wisconsin who was working on another kind of accelerator called a betatron. In 1942, Kerst published a paper about the betatron, and Grimmett wrote him asking for further details about it. In his reply, Kerst had written: "...I hope that some day you can also find betatron helpful." [56].

Grimmett was also aware of the development with linear accelerators for radiotherapy use, which resulted from the work on radar. Radar, an acronym for "radio detection and ranging" was one of the powerful developments to emerge from World War II. It was based on microwave technology. In the United States, the Varian brothers had developed the klystron, a high-powered microwave amplifier, and in the UK,. the magnetron, a lower-power microwave source and amplifier had been developed by Randal and Boot. It was quickly realized that such devices could be used to accelerate electrons to high, megavoltage, energies which could then be used to generate megavoltage X-ray beams. Another of the notes, he wrote in his diary on the meeting of the HPA in London in January 1948 concerned the status of linear accelerators for radiotherapy given by Mr. Newberry of the General Electric Company (GEC) of England.

Linear Accelerators
  4–5 MeV certain in 1 m
  Probably more mobile then e.s. generator at these voltages
  But cumbersome at 20 MeV
  Beam very narrow at 20 MeV
  Wobbulate (sic) patient on table to cover field… Probably find 5 MeV optimum.

(Wobbulate seems to be a word that Grimmett invented to indicate that with a narrow X-ray beam, the patient would have to be moved back and forth (i.e., wobbled back and forth) under the beam to adequately cover the area that needed to be treated.)

So in addition to cobalt-60 machines, Van De Graaff accelerators, betatrons and linear accelerators were also being developed at the same time. And they were all used for radiotherapy. The Van de Graaff units, which were quite large and operated at a few Mev, were in direct competition with the cobalt units and were not extensively used, although the British had one at the National Physical Laboratory in their radiation standards laboratory. Betatrons enjoyed a fair period of popularity in the 1950s and 1960s. They generally operated at higher energies in the range approximately 20–40 MeV and so met a need for radiotherapist who wanted the higher energies. Watson and Johns installed a 22 MeV betatron in Saskatoon in 1949, and at the same time, they were developing their cobalt unit. Grimmett and Clark had a number of discussions concerning a betatron for MDAH, and Grimmett designed the space and viewing window for it in the new hospital (see Chap. 9). One was purchased and installed in 1954 when the new hospital was completed. Ironically, when the Grimmett cobalt unit was decommissioned in 1963, it was replaced with a small betatron. However, betatrons were large, bulky and noisy and the output was not particularly high so treatment times could be fairly long. The 1976 IAEA Directory of High-Energy Radiotherapy Centers mentioned above gave the following numbers: there were 24 Van de Graaffs and 219 betatrons in use worldwide for radiotherapy, representing 1 and 7 %, respectively, of the total machines in use. The same percentages applied to the United States.

The first medical linear accelerator was installed at Hammersmith Hospital, London, in 1952, and therapy began in August 1952. It will be recalled from Chap. 4 that Hammersmith Hospital was the last place in the UK where Grimmett worked as a medical physicist for the MRC before his difficulties with the MRC and his fellow workers. When Grimmett moved there with the Radium Beam Therapy Research unit in 1941–1942, Hammersmith Hospital became established as a leader in radiotherapy and medical physics in the UK. By 1955, the UK had three 4 MV and one 8.5 MV linear accelerators in clinical use.

In 1947, in California, Varian Associates was formed to manufacture klystrons for radar applications, but they soon entered the field of linear accelerators called linacs for cancer radiotherapy. Today, they supply most of the world's clinical linear accelerators. One of their early accelerators went to the Stanford Medical Center in California where the first patient was treated in January 1956. Grimmett's insight into linear accelerators was quite acute indicating that he had a clear grasp on what was needed for radiotherapy treatment units. Although the early machines required an accelerator tube of about a meter for 4–6 MeV, later developments in accelerator technology made it much shorter so that linear accelerators in that energy range are now very compact. He was right in realizing that linacs in the 20 MeV range would be bulky. He knew that all X-ray beams from linacs would be peaked in the forward direction and more so at 20 MeV making the beam narrow. Putting a shaped attenuator in the beam, called, a flattening filter solved this problem. This technique was used for all the megavoltage therapy machines, except the cobalt units.

In linear accelerators, high-energy electrons hitting a metal target produce the X-ray beam. As the electrons loose energy, X-rays are produced. The diameter of the electron beam hitting the target, called the spot size, is equivalent to the source diameter of the cobalt units.

For linear accelerators, the spot size is approximately 3 mm and as discussed above the diameter of a cobalt source is 1.5–2 cm. In addition, the distance of the target from the isocenter for the linacs became standardized at one meter; the distance from the source to isocenter for cobalt units was 80 cm. These two factors, the small spot size and the greater treatment distance, resulted in well-defined treatment beams with small penumbras (less than 5 mm), as compared to cobalt-60 beams (more than 1 cm). In addition, the beam flatteners produced a more uniform beam across the face of the beam, whereas, the cobalt beam is rounded. These factors combine to allow better precision in dose delivery for the linacs and this advantage came at a time when radiotherapists were seeking greater precision to improve their treatment results. Add to this that the output for the linacs was higher, meaning shorter treatment times, there was no problem with source decay and adjusting the output monthly, no problems with source disposal or the cost of source replacement (although magnetrons and klystrons do have a finite life-time and are costly to replace). As linear accelerator technology improved, reliability approached that of cobalt units. And linear accelerators are much more adaptable to computer control, which is now central to advanced radiotherapy.

So comparing the cobalt unit with a similar energy 6-MeV linear accelerator, the advantage swung to the linear accelerator, and in the United States, the use of cobalt-60 units peaked in 1985 and was overtaken by linear accelerators. (In addition, there are dual-energy linear accelerators with a low- and high-energy X-ray beams, and with the option of also using electrons for radiotherapy makes the equipment very versatile and cost efficient).

The exact number of cobalt-60 machines in use worldwide is difficult to come by since they are now used in over 100 countries. UNSCEAR in a survey of radiotherapy equipment for the years 1991–1996 reported 2576 cobalt-60 units versus 4239 linear accelerators worldwide [198]. In 2012, IAEA reported that there were 1625 cobalt units and 8481 linear accelerators [199]. For the United States, the use of cobalt-60 units peaked in the mid1980s at around 1,000 units, and in the UNSCEAR survey for 1991–1996, the number was down to 504 and is even lower in 2012.

Never the less, there are still 1,000–2,000 cobalt-60 units in use worldwide especially in developing countries where cobalt units are much easier to use and maintain than linear accelerators. Grimmett's dream that cobalt-60... "would seem to be a sound way of using atomic products, which would bring the benefits of high-voltage radiation within the reach of ordinary hospital" [120], has been fully justified. Millions of cancer patients cured by their cobalt-60 treatments would agree.

Cobalt-60 was not only considered for a replacement for radium in teletherapy machines it was also considered as a replacement for radium needles and tubes. The Science News-Letter, May 1 1948, put it this was:

> The material (radioactive cobalt) will cost about a tenth of what radium costs... The total would probably come to between $60 and $75. The cost of an equivalent amount of radium, on a dosage basis would be $500...
>
> Radioactive cobalt would be used in needles or tubes in the same way that radium is used for cancer treatments [16].

Fletcher and Grimmett also proposed research programs to look into these uses.

Radium needles were platinum-iridium hollow needles (a few centimeters in length and a millimeter in diameter) containing a few milligrams of radium which were surgically implanted into tumors, left in place for several days and then removed. Head and neck tumors were often treated this way. The first step in the decay of radium is into radioactive radon gas before going through an additional 12 steps before finally becoming a stable isotope of lead, with a half-life of 1,602 years. Since no radon can be allowed to leak out, the radium was hermetically sealed in platinum-iridium needles. Since helium is also a by-product of radium decay (the alpha particles emitted during decay are just the nucleus of the helium atom), the integrity of the seal had to be maintained as pressure built up on the inside. The needles went through traumatic handling during the implantations. The needles would become bent and sometimes cracked, and a routine testing program was required to ensure that they were safe to use. To quote the Science News-Letter again, "Radium, because of the radon gas which emanates from it, involves a more difficult handling problem (Than cobalt-60)" [16]. Several studies were undertaken using cobalt-60 in the place of radium in needles [200], but it was not a successful substitution. Eventually, the radioactive isotope that replaced radium needles was iridium-192 (the same isotope used in Freundlich's teletherapy unit) in the form of radioactive wire, which could be quite easily implanted into tumors.

Grimmett and Fletcher certainly intended to use cobalt-60 to replace radium tubes. The radium tubes where also made of platinum and iridium, 20 mm long and 3 mm diameter containing a few milligrams of radium hermetically sealed inside. Again the integrity of the seal had to be routinely checked. The radium tubes were used with a variety of applicators and placed in various body cavities where there was cancer. They could also be put in molds and placed on a patient's skin to treat surface cancers. One of the primary uses was in the treatment of cancer of the cervix, and Fletcher and Grimmett started out to design an applicator to treat cervix cancer using cobalt tubes rather than radium ones, but the cobalt tubes never worked out and the applicators which were very successful used the radium tubes [201]. As with the radium needles, however, radium tubes were eventually replaced with radioactive cesium-137 tubes.

Radium was used in just one other way to treat cancer. The radon gas from a radium solution could be captured and sealed in very small glass containers a few millimeters in diameter, each containing a few millicuries of radon, called radon seeds. These radon seeds were implanted directly into tumors, and since the half-life of radon is 3.8 days, they were left in place. The isotope replacement for radon became radioactive gold-198, which has a half-life of 2.696 days. Small radioactive gold seeds can be implanted directly into a tumor and left in place.

Cobalt-60 has found applications other than medical ones, including being used as the source of radiation in sterilization plants for medical supplies and food, etc. It is also used in small irradiators for research purposes.

Finally, there is one other use of cobalt-60 in radiation therapy, which is of great importance. When cobalt-60 units were introduced into use, the question of calibrating their output arose. For the kilovoltage X-ray machines then in use, the output was measured in roentgen per minute or r/min using small ionization chambers. These ionization chambers had to be calibrated against a national standard. For the United States, this was done at the National Bureau of Standards (NBS) in Washington D.C. (Now the National Institute of Standards and Technology, NIST in Gaithersburg, Md.). And the standard was maintained with "free-air standard ionization chambers." The quantity used to express the output of the X-ray machines was exposure in units of r/min. But the γ-rays from cobalt-60 are too high in energy for such chambers to work, and another approach was used. This made use of the Bragg-Gray cavity theory from which the exposure can be calculated. But to use the Bragg-Gray theory, the volume of air in the chamber and, therefore, the mass of air in the chamber must be known very accurately. This was almost impossible to do for the small ionization chambers used to calibrate local cobalt units. NBS constructed a set of precision-made ionization chambers for which the volumes where accurately known and used the Bragg-Gray cavity theory to calculated exposure, in r/min, at the accuracy required for a standards laboratory for their cobalt beam. It was against this standard that physicists could have their ionization chambers calibrated. In effect, the calibration of the local physicist's ion chamber in the standards laboratory cobalt beam was equivalent to determining the volume of the ion chamber so that the Bragg-Gray cavity theory could be used. This approach is the basis of many national and international radiotherapy beam calibration protocols in use today. The calibration is now more often in terms of absorbed dose in units of gray/min and can be applied to the calibration of the high-energy beams of different energies from linear accelerators and also to electron beams [202].

Even as the use of cobalt-60 units decrease, the majority of the world's radiotherapy is now done on linear accelerators calibrated with ionization chambers standardized in a cobalt-60 beam. The Accredited Dosimetry Calibration

Laboratory (ADCL), in the department that Grimmett established, uses one of the remaining cobalt units at MDACC for this purpose. Grimmett who had planned to build his own free-air standard ionization chamber for ionization chamber calibration would be pleased.

Chapter 12
Erratum to: Cobalt Blues

Peter R. Almond

Erratum to:
P. R. Almond, Cobalt Blues, DOI 10.1007/978-1-4614-4924-9

Author Biography

Peter. R. Almond, Ph.D. received his undergraduate honors degree in physics from Nottingham University and his training in medical physics from Bristol University, in the United Kingdom. In 1959 he moved to the United States as a fellow in Medical Physics at the University of Texas M.D. Anderson Hospital and Tumor Institute earning his master's and doctoral degrees in nuclear physics from Rice University in Houston in 1960 and 1965 respectively. From 1964 to 1985 he worked at The University of Texas M.D. Anderson Cancer Center in Houston in the Physics Department, where he served as the Head of the Radiation Physics section and director of the Cyclotron

---

The online version of the original book can be found under DOI 10.1007/978-1-4614-4924-9

P. R. Almond (✉)
Calle Ronda Place 203, Houston 77007-1155, USA
e-mail: palmond@mdanderson.org

Unit and Professor of Biophysics. He was a member The University of Texas Graduate School of Biomedical Sciences from1966 to 1985 and from 1999 to 2012 as a Distinguished Senior Lecturer. From 1985 to 1998 he was Vice-Chairman of the Department of Radiation Oncology at the University of Louisville.

Dr. Almond has helped to develop cancer treatments with various forms of radiation, including high-energy photons, electrons, and neutrons. He has also been instrumental in developing basic measurement techniques for these radiations—writing calibration protocols for the United States, the IAEA, and ICRU. He helped found the journal Medical Physics, has served as the North American Editor for Physics in Medicine and Biology, and was the Founding Editor and Editor-in-Chief of the electronic Journal of Applied Clinical Medical Physics. He has served on numerous national and international committees and councils including the NCRP, the NRC, and the NIH on the Radiation Study Section, serving as chairman for 2 years. He worked on the dosimetry for the atomic bomb survivors for the NAS/NRC.

Academically, Dr. Almond has supervised over 25 masters and doctoral students in medical physics. He has authored or co-authored over 100 scientific articles and numerous chapters in radiotherapy textbooks. He has served as President of the American Association of Physicists in Medicine and as Chairman of the Board of Chancellors of the American College of Medical Physics. Dr. Almond is a fellow of the AAPM, ACMP, ACR and the IOP.He is diplomat of the American Board of Radiology and the American Board of Medical Physics. He is licensed as a Professional Medical Physicist in the state of Texas and a Chartered Physicist (C Phys) in the United Kingdom. He has received the Coolidge Award and the Marvin M.D. Williams Professional Achievement Award from the AAPM and the ACMP respectively. He has twice received the Farrington Daniel Award for the best scientific paper on Radiation Dosimetry in Medical Physics. Now retired, he works on the history of medical physics.

Epilogue
Grimmett the Man

Like most of us Grimmett was a complex person. Those that knew him talked about how kind and thoughtful he was, what an English gentleman he was and how much they liked him. On the other hand, there were those who could not get along with him and disliked him.

He came from a poor home in a working class section of London at a time when class distinction was deeply ingrained in English life. To get ahead in life, he had to do it on his own. Life was very competitive in his home with three brothers and no sisters; as the eldest he would have had to set the example. They were all musical and this would have increased the competitiveness. His education was only affordable because he obtained scholarships and took nighttime jobs. He worked hard to improve himself and be accepted. He became such a good pianist that for a short period of time he earned his living by playing the piano on steamship liners going to South America and for some months in South America. When he returned to England, he played for silent movies and theaters.

He was a precise and neat person and was determined that his work reflected who he was. His laboratory notebook is written in a clear legible hand, the diagrams beautifully drawn and the data entered in precise order. He required that any equipment that he designed and built or had built was not only functionally but also esthetically pleasing to look at. His reports on meetings, trips and conferences were well written leaving the impression that in reading them he had conveyed the essential facts of what had taken place. Without them much of what is in this book could not have been written.

He was a man of many interests, scriptwriter, flying, bookbinding, calligraphy, jewelry and a worker in precious metals besides his music. He was known to grow star sapphires and had a small workshop in his home in London and had a jewelry business on the side with his brother Rubin.

John Reed who probably knew him better than any of his contemporaries in England said that he, "…was kind, gentle, always soft spoken and quite imperturbable." He had compassion and as Reed put it could be, "…filled with distress," at the sight of out-of-work musicians because he had played with them and who were out of work because of the "talkies" as he had been. John Read's obituary in the British Journal of Radiology, September 1951, recalls:

> Grimmett showed that with few initial advantages other than ability, courage and tenacity, he could be a twentieth-century pioneer and adventurer, yet without aggression or acquisitiveness, laying the foundations of an expanded application of physics in the cure of disease. [3]

Those that worked with him and for him were extremely loyal to him. The sketch presented to him when he left Paris sums it up; even the Eiffel Tower is in tears to see him go. Scientist and musician Dr. Grimmett was a unique personality. He was charming and delightful and enjoyed widespread admiration and respect.

J.E. Roberts who was a contemporary medical physicist in London and was at The Cancer Hospital (Free) from 1932 to 1937 and who along with Grimmett and others was a founding member of the Hospital Physicists' Association called him, "assuredly one of the characters of early medical physics."

He wrote of Grimmett:

> Although he made some valuable contributions to radiological physics, particularly in instrumentation, he would probably have claimed that his greatest contribution to human welfare and happiness was as a "pop" musician and particularly as a pianist. [203]

He may have played "pop" music when he performed in the theaters during his university days and for the passengers on the liners to South American, but he was also a classicist. In January 1951 Ann Holmes, the fine arts editor of the Houston Chronicle, interviewed him prior to the first, and as it turned out his only piano concert, he was to present in Houston. Her article appeared in the January 25 edition of the Chronicle under the headline, "Physicist Grimmett Is Successful Musician Also." She noted that he was a quiet man given to understatement. She reported that he had started playing very early in life and studied for many years in London and served as an accompanist for a number of leading singers and instrumentalists in London. It was with small chamber groups in the repertory theaters of Hampstead that she said he described as the 'arty' section of London which had had a special appeal to him. His program, she wrote, was of impressive dimensions and included Scarlatti sonatas, Schubert Impromptus, Chopin and Beethoven works. There was no "pop" music [204].

The concert he gave on January 30, 1950, at 8 p.m. in the Carter Recital Hall, advertised as "The finest studio auditorium in the southwest," was purely classical music. It was a private recital for the staff of the M.D. Anderson Hospital and their friends with the proceeds going to the hospital's patient welfare fund. He played four sonatas by Scarlatti, an Impromptu by Franz Schubert and a Sonata by Beethoven. After an intermission, he played Pictures at an Exhibition by Mussorgsky for which he wrote the program notes, and after another intermission, he finished the concert with two Preludes by Chopin and the Hungarian Rhapsody by Frantz Liszt [205]. No record has been found of the amount of money that was raised.

His hope had been that by coming to Houston he would be able to reestablish himself as a medical physicist to build up a name for himself and his department. But he did not have time.

Reed wrote for Grimmett's obituary in the British Journal of Radiology:

Then in 1944 he disappeared from his usual scenes, reappearing briefly from time to time from Paris, or wherever his service for the British Council, UNESCO, and UNO had taken him. Finally it was reported that he had settled in Houston, Texas, where he was developing a radiobiological research laboratory in the really big way which suited his temperament [3].

But Grimmett could be stubborn and obstinate and very defensive about his work. Perhaps, the first instance of this was his disagreement with Mayneord over the report on Radium Beam Therapy Research 1934–1937. In reading the report, it is hard to understand why Mayneord was upset, but he was involved with Grimmett in the dosimetry of the radium teletherapy units and there was an agreement with the Royal Cancer Hospital that they would review manuscripts prior to publication. When Grimmett became aware of Mayneord's concerns, he did make a concerted effort to address them and suggested a follow-up publication, but when Mayneord was still not satisfied, Grimmett dug his heals in and cancelled the idea of a joint publication. Mayneord was not only a leading figure in British medical physics he was extremely well connected in London, and it could not have been too helpful to Grimmett's career to have crossed swords with him.

At times he was perhaps too disconnected from others to realize what their impression of him might be. This seems to be the case when he was let go from the Medical Research Council. Although he worked hard and often long hours, it was on his schedule and not other peoples' so when he was not around during their working hours, he was seen as lazy and uninterested. When this was pointed out to him, he seemed surprised and hurt and defensive. Perhaps, this meant that any relationship would eventually deteriorate.

This was true of Grimmett's relationship with Dr. Gilbert Fletcher during the time they worked together. In the memo that Grimmett sent to Fletcher on April 4, 1950, his annoyance and disgust with Fletcher clearly comes through. Fletcher could be very difficult to work with and provoked strong reactions from people, and the relationship between the two would never be restored. Grimmett was proud of his work, his accomplishments and his ability to solve problems that were presented to him. He knew what he had done and what he could do and he did not appreciate anyone taking credit or trying to take credit for what he had done. The M.D. Anderson Hospital position represented a chance to re-establish his medical physics career, an opportunity he stated repeatedly in his letters to his wife back in England before she joined him in the spring of 1949. His appointment as a Fellow in the Institute of Physics in 1946 he regarded as recognition of him as a physicist, and he made sure that a notation about it was put in his files at UNESCO.

But he did not have enough time to establish himself internationally. Although Grimmett came up with the concept of the cobalt-60 teletherapy unit, he died before his unit could be put into use. The Canadians who independently developed the concept and put it into clinical use received most of the credit. J.E. Roberts writing his memoirs in the late 1990s never mentions Grimmett's contribution to the cobalt-60 units. He ends his comments on Grimmett by writing, "He ended up as a physicist at the M.D. Anderson Hospital in Houston, Texas." [203] There is not even the recognition that Grimmett was the chairman of his own independent department at the hospital. Grimmett would have been disappointed.

Appendix A
Principles of Radiotherapy

The basis for radiotherapy is that ionizing radiation destroys cancer cells. Ionizing radiation has the ability, when it interacts with matter, to set free some of the electrons, associated with the atoms of the material, which allows them to move through the material being irradiated. The electrons have a negative electric charge leaving the remaining atoms with a positive charge, creating what is called an ion pair. Hence, the term "ionizing radiation." As the electrons move through the material, they can create further ion pairs. It is these negative and positive ions that have a biological effect. For clinical purposes, the energy of the radiation must be greater than that of ultraviolet light, not only to have enough energy to create ion pairs but also to have enough energy to penetrate into the tissue. There are two sources of ionizing radiation, radioactive materials and radiation producing machines such as X-ray machines or linear accelerators. Radioactive materials emit ionizing radiation in the form of alpha rays, beta rays and gamma rays, and in general, it is the gamma rays that are used for treatment purposes because of their ability to penetrate into the tissue. X-rays and gamma rays are, from a physics viewpoint, identical, and they are both electromagnetic radiation. Gamma rays are emitted by radioactive materials, while X-rays are produced when high-energy electrons hit a target (in an X-ray tube) and are stopped. The energy of the X-rays can never exceed the energy of the electrons producing them, and because there is a range of X-ray energies produced, the average energy of the X-rays is approximately half that of the electron energy. Gamma rays have single energies.

To match the average energy of the X-rays to the gamma ray energy, the energy rating of the X-ray tube must be about twice the energy of the gamma rays. In the middle of the twentieth century, it was normal for electromagnetic radiation to be characterized by the wavelength of the radiation. The product of the radiations wavelength with its frequency gives the velocity of the radiation. Since all electromagnetic radiation (including X-rays and gamma rays) travels at the speed of light, which is constant, the shorter the wavelength, the higher the frequency, and since frequency is directly related to energy, the shorter the wavelength, the higher the energy. When Grimmett became a medical physicist in the 1920s, X-ray energies were designated by their wavelength. Today, it is customary to express the energy of ionizing radiation directly in terms of electron volts (eV). For therapy X-rays with energies of thousands of electron volts (keV) or higher are required; gamma rays from radioactive materials have energies in the millions of electron volts (MeV). Grimmett understood that to match the 1.25 MeV gamma ray energy of cobalt-60 would require a 3 MeV X-ray tube.[1]

Radiotherapy was the term used for much of the twentieth century, and the doctors who practiced it were generally radiologists located in radiology departments in hospitals. Radiologist practiced both diagnostic and therapeutic radiology. If a doctor specialized in treatments only, they were called radiotherapists. In the twenty-first century they are more generally referred to as radiation oncologist and practice in departments of radiation oncology, separate from departments of diagnostic or imaging radiology.

The French physicists Pierre and Marie Curie at the end of the nineteenth century and the beginning of the twentieth century discovered radium which is radioactive. They soon noticed that the application of radium to the skin had a biological effect and radium was quickly made available to the medical profession to treat cancer. However, ionizing radiation also destroys normal human tissue cells, so that in treating cancer with radiation, care must be taken to get as much of the radiation to the cancer and as little a possible to normal tissues. Milligram amounts of radium were therefore put into small metal containers, about a millimeter in diameter and about a centimeter in length and made into tubes and needles. If cancer was in or near a body cavity, the tubes would be put into the cavity. For other anatomical sites, needles could be inserted directly into the tumor. For skin cancer the radium was put into plaques that had been molded over the skin and designed to hold the radium close to the lesion. In this way the radiation dose to the cancer was maximized, and the dose to surrounding normal tissue was kept to a minimum. In general, one or two applications of the radium were made and in many cases were successful in curing the patient or controlling the disease. Radon, the radioactive gas emitted by radium, was also put into small capsules and used for therapy. This approach was not without its hazards, however. The doctor had to handle the radium when inserting it into the patient and over time could receive a significant radiation dose that could and often did result in damage to the physicians' fingers and the development of cancer. The needles and tubes of radium could also leak radiation if not handled properly, which could cause radiation contamination in the hospital. This type of cancer treatment was called by many names such as radium interstitial therapy, mold therapy, intracavitary therapy or more generally radium therapy. The overall term that came to be used to describe them all was brachytherapy, meaning therapy at a short

---

[1] With the development of linear accelerators in particular, it became possible to build X-ray generators in the multi- megavolt range, and by the end of the twentieth century, linear accelerator x-ray machines had all but replaced the cobalt-60 units.

distance since the radium was placed next to, in contact with, or at a short distance from the cancer. Because interstitial radium therapy required placing needles in the patient under anesthesia, surgeons often practiced this kind of radiotherapy.

At the same time that radium was discovered, Wilhelm Roentgen discovered X-rays in Würsburg, Germany. The biological effects of X-rays were also noted right away, and doctors started treating cancer with X-rays. Because of the size of the X-ray tubes, the patient was placed some distance from the X-ray source and the X-rays were directed to the site on the patient that needed treatment. It too had its hazards and in the early days, the operators of the X-ray equipment also received large amounts of radiation. It was soon realized that the X-ray tubes had to be adequately shielded and the X-rays had to be restricted, by collimation, to a beam just large enough to irradiate the cancer and avoid as much normal tissue as possible. This type of treatment became known as external beam treatment, X-ray treatment or more generally as teletherapy, meaning treatment at a distance. Its major benefit over brachytherapy was that it required no surgical intervention.

Although brachytherapy was generally given in one or two treatment sessions, each lasting hours or sometimes days, it was determined that teletherapy was best given on a daily basis with treatment times of a few minutes each day and the course of treatments lasting several weeks. This type of treatment became known as fractionated treatment. The break between each daily treatment gave time for the normal tissues to recover better, and the accumulated dose of radiation to the cancer could be increased to high enough levels to kill all the cancer cells.

One of the questions in the first half of the twentieth century for radiotherapy was the following: Could the benefits of external treatments be realized using radium as the source of the external radiation? Since the gamma rays from the radium and the X-rays from the X-ray tube were known to be identical from a physics viewpoint (they are both electromagnetic radiation) differing only in energy, could the X-ray tube be replaced by a sufficient amount of radium to give treatments at a distance? If it could, then high-voltage electrical equipment, in the range of hundreds of kilovolts, and its associated electrical hazards could be eliminated. There would also be no need to replace costly X-ray tubes that had finite lifetimes. The half-life of radium is long, 1,600 years, so that its output was considered constant with time and would never need replacing. There was also a belief among some radiologists that gamma rays from radium were medically superior to X-rays because of their shorter wavelength, and therefore higher energy. Gamma rays from radium have energies up to 1–2 MeV. To match this would require an X-ray tube of 3 MeV, and to build and operate such an X-ray tube did not appear feasible at the time. There were good reasons, therefore, to see whether radium could be used to replace X-ray tubes. But there were serious problems. It required a large amount of radium, which was very expensive (approximately \$750 per milligram in today's dollars), and it was generally considered that 4 g of radium was the minimal amount needed (i.e., a \$3 million investment). Only a few places could consider this kind of expense. The use of such large amounts of radium meant that the apparatus holding the radium source had to provide an adequate degree of protection for the people unsung the equipment. The containers therefore had thick walls of dense metal to absorb the radiation, and an aperture to allow the exit of the beam of gamma rays for the treatment. Because of their shape, size and the fact they were made of metal, they were often called radium bombs. Single sources of 4 g of radium were not easy to come by, and the sources in the radium bombs consisted of multiple sources of lower amounts. For example, twenty tubes of 200 mg each might be used. This meant that the size of the radium source in the radium bomb was quite large and considerable self-absorption of the gamma rays in the source itself took place, making the source strength effectively much less. Even with 4 g, the radiation output of these units was not high, and in order to give the treatments, the source had to be close to the patient surface. In most cases between 5 and 10 cm (compared with 75 cm for X-ray treatments), even so the treatment could last up to 30 min or longer. But a 5–10 cm treatment distance counteracted the very characteristic for which radium might be used in the first place.

For external beam treatments, the radiation must pass through the skin of the patient to reach the cancer below the surface. The essential problem of radiotherapy at the time therefore was to deliver, at depth in tissue, as high a percentage as possible of the dose received by the skin. The reason being that the skin reaction was the limiting factor as to how much dose could be given. The radiation would produce a sunburn-like reaction, and there are limits in the amount of radiation to the skin beyond which permanent damage would be done.

All other factors being equal the penetration of the radiation into the body is determined by the energy of the radiation and the distance of the source of radiation from the skin surface (called the treatment distance). The higher the energy, the more the radiation penetrates, but it is also dependent upon the inverse square law, the intensity drops off as one over the square of the distance from the source. If the distance from the source is doubled, the intensity is decreased by a factor of four, for example. If the treatment distance is short, of the order of the depth into the body to which the radiation needs to penetrate, the inverse square law predominates and the energy has little effect. On the other hand, if the treatment distance is long compared to the depth of interest, the energy of the radiation predominates.

For radium bombs that had a treatment distance of 5–10 cm, the penetration of the radiation into the body (to depths of say 10 cm) would be almost identical to that of a kilovoltage X-rays at the same treatment distance and there would be no physical advantage to using the radium with its higher-energy radiation, except if there was a biological advantage to the higher-energy radiation as some radiologist believed.

Figure A.1 shows the measured depth doses of 200 kv. X-rays at a 50 cm. focus-skin distance (curve B), the depth dose calculated for a hypothetical radium unit utilizing the same focus-skin distance (curve A) and the depth dose for an existing 4 g unit in which the focus-skin distance was 8.0 cm (curve C) [47]. Comparison of the curves A and C shows the gain that would have been achieved

Appendix A: Principles of Radiotherapy

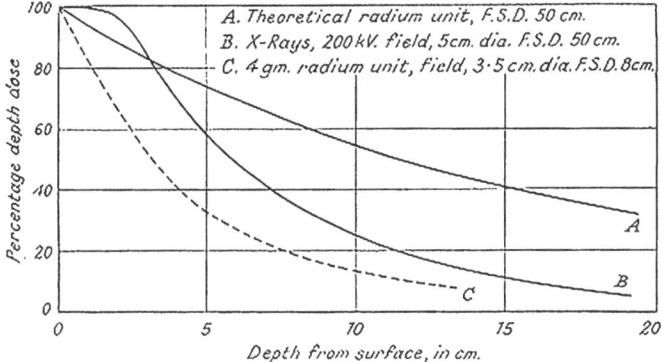

Fig. A.1 Comparison of the depth doses from (*A*) a theoretical radium unit having 50 cm F.S.D., (*B*) 200 kv. X-rays, 50 cm. F.S.D., (*C*) an actual radium unit, 8 cm. F.S.D. [47]

if such a unit could have been built and how much better it would have been compared to the X-rays (curve B)

But with the focus-skin distance at 50 cm compared to 8.0 cm, the intensity of the radiation at the surface would be reduced by almost a factor of 40. To achieve the same dose rates therefore with the longer treatment distance machine as with the shorter treatment distance machine (whose treatment times of 30 min or so were too long and not ideal) would require approximately 160 g of radium, which was not possible from a cost, safety or even availability considerations. It would seem that the radium teletherapy units were stuck at treatment distances of 10 cm or less.

X-ray units did not have this problem. Even at the extended treatment distances of 50–75 cm, the output was high enough that treatment times of a few minutes were possible.

The other factor that came into play was the penumbra of the beam, that is, the fall off of the dose, at the edge of the beam. This is dependent upon the geometry of the treatment machine. With short treatment distances and large source sizes, the penumbra is large, which was the case with radium units and was not desirable to radiation treatments. For X-ray units at longer treatment distances and very small source size, penumbra becomes almost nonexistent, which is a large advantage for X-rays.

It is surprising therefore that with these three major disadvantages, poor depth of penetration, poor penumbra and low output that radium teletherapy showed any promise at all. It succeeded to the extent that it did because the main area in which it was tried was for head and neck cancer where penetration is not a big issue, a large penumbra can be tolerated and immobilization of the patients head can be used to minimize the effect of the long treatment times.

Appendix B
Principles of Reactor Production of Cobalt-60

The story of the completion of MDAH/ORINS cobalt-60 unit is the story of two nuclear reactors, the Oak Ridge graphite reactor in Oak Ridge Tennessee and the Canadian heavy water reactor at Chalk River in Ontario Canada.

Cobalt exists naturally only as non-radioactive cobalt-59 with 27 protons and 32 neutrons in its nucleus. The addition of one extra neutron in the nucleus creates radioactive cobalt-60, but does not change the chemical characteristics of cobalt. Placing cobalt-59 into an intense field of slow or low-energy neutrons such as found in the interior of a nuclear reactor can produce cobalt-60.

In a reactor the fission of uranium-235 is initiated when it captures a slow (low energy) neutron. In the process heat and a number of high-energy neutrons are released. The coolant carries off the heat. The high-energy neutrons are slowed down by the moderator and can then be captured by other atoms of uranium-235, causing further fission.

As the process repeats itself, a chain reaction is produced. Control rods of boron or cadmium, which are proficient at absorbing the neutrons, can be inserted into or removed from the reactor in order to adjust the power level or to shut the reactor down. Samples, such as cobalt-59, can be inserted into the reactor to be activated as long as they do not absorb too many neutrons and reduce the power of the reactor to unacceptable levels.

The Oak Ridge reactor was a 1,000 kW, carbon-moderated, air-cooled reactor built in 1942–1943 as a pilot plant, to demonstrate the feasibility of producing plutonium from uranium in large-scale production units, and partly to provide plutonium that was badly needed for experimental purposes. It was considered the first milestone in the creation of the atomic bomb that ended World War II. The moderator was a cube of graphite, 7.3 m (24 ft) on each side, as the moderator with tubes containing the uranium fuel in a horizontal matrix running through the moderator. Vertical boron steel control rods could be moved in and out of the moderator to control the reactor. There were a number of horizontal channels, at right angles to the fuel elements, in the moderator, into which long graphite holders or stringers could be inserted. The stringers contained cylindrical holes into which gas-tight aluminum casings could be inserted. Normally uranium would be put into the casings for plutonium production, but the same type of system was used to place isotopes to be activated. For nearly 20 years, it was one of the world's foremost sources of radioisotopes for medicine, agriculture, industry and research. In the early 1950s, however, the top priority was for defense work, and the production of radioactive isotopes for non-military purposes came second. There was always a trade-off between the number of samples that could be put into the reactor and the power level of the reactor; too many samples would absorb too many neutrons, and the power level would drop (Fig. B.1).

Although quite a few cobalt-59 samples were placed in and around the Oak Ridge reactor, they were generally in areas of low neutron flux, so that it took a long time to activate the samples to acceptable levels of activity. The Oak Ridge reactor was, therefore, not ideal for activating kilocuries of cobalt-60 to high specific activities. It was into this reactor, however, that the initial cobalt sources for the ORINS/MDAH cobalt unit were placed.

Decommissioned in 1963, the Graphite Reactor is now a National Historic Landmark.

The reactor at Chalk River, Ontario, Canada, on the other hand provided a neutron flux many times greater (approximately 100 times greater than the Oak Ridge reactor) than any other reactor at the time. Called the NRX it was a heavy water-moderated, light water-cooled reactor. NRX was for a time the world's most powerful research reactor. It was a cooperative effort between Britain, the United States and Canada during World War II. NRX was a multipurpose research reactor used to develop new isotopes, test materials and fuels and produce beams of neutrons.

In a heavy water-moderated reactor either inserting the control rods or removing the heavy water moderator can stop the reaction.

The NRX reactor incorporated a sealed vertical aluminum cylindrical vessel which held 14,000 L of heavy water and helium gas and about 175 six centimeter diameter vertical tubes in a hexagonal lattice. The level of water in the reactor could be adjusted to help set the power level. Sitting in the vertical tubes and surrounded by air were uranium fuel elements or experimental items, cooled by light water.

Twelve of the vertical tubes contained control rods made of boron powder inside steel tubes. These could be raised and lowered to control the reaction, with seven inserted being enough to absorb sufficient neutrons that no chain reaction could happen. The reactor began operation on July 22, 1947, under the National Research Council of Canada and was taken over by Atomic Energy of Canada Limited (AECL) in late 1952. It operated for 45 years, being shut down permanently in 1992.

Because of its higher neutron flux and the number of tubes available for samples to be activated, the NRX was far superior to the Oak Ridge reactor for producing large curie amounts of radioactive cobalt-60 with high specific activity.

The activation of a sample depends upon the neutron flux in the reactor, the probability of the target nucleus absorbing the neutron (known as the cross section

Appendix B: Principles of Reactor Production of Cobalt-60

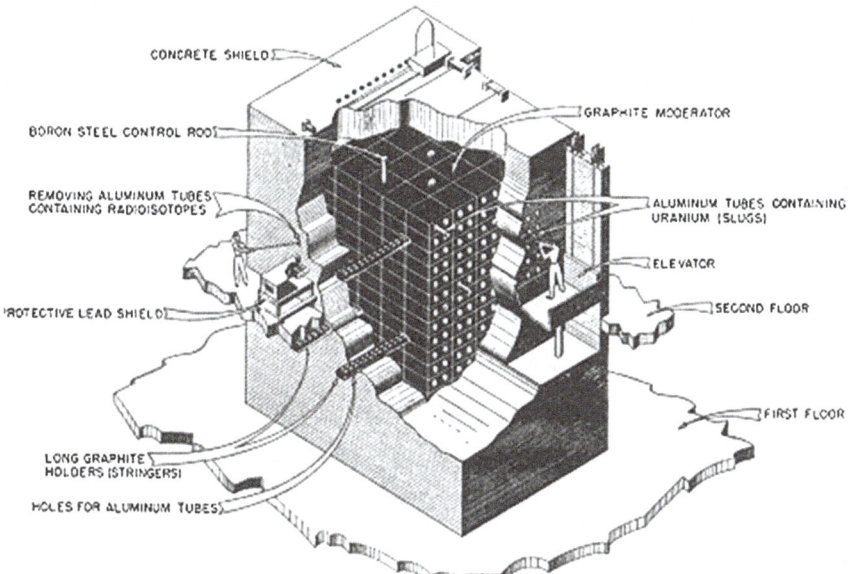

Fig. B.1 Cut-away view of the Oak Ridge reactor [206]

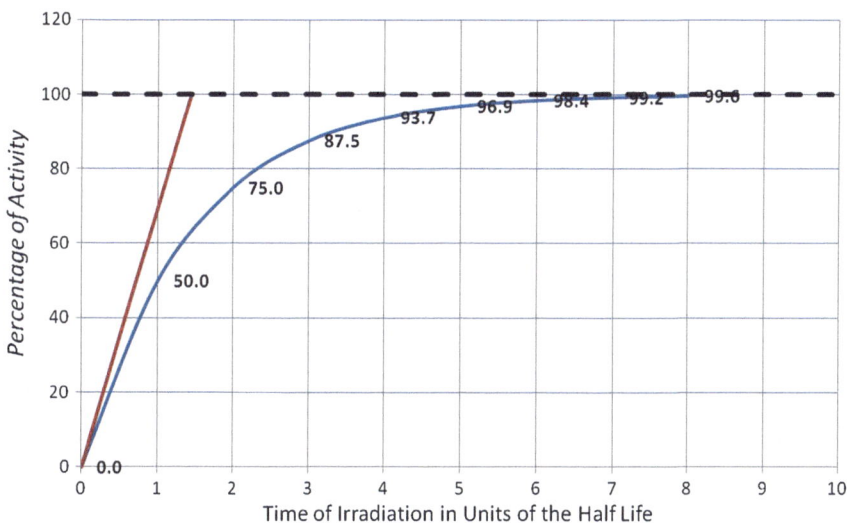

Fig. B.2 Percentage buildup of activity in a reactor versus the irradiation time in units of half-life

for the interaction), the half-life of the isotope being produced and the amount of the sample being activated but not the shape of the sample. The ORINS/MDAH source consisted of four plaques of cobalt, 2 × 2 × 0.25 cm. Since a total activity of 1,250 Ci of cobalt-60 was needed, each source had to be activated to 312.5 Ci.

In the Oak Ridge reactor, it would have taken nearly 7 years to reach that level. In the Chalk River reactor with a hundred times higher neutron flux, it would take 10 months (Fig. B.2).

The sources for the ORINS/MDAH cobalt irradiator were eventually removed, therefore, from the Oak Ridge reactor and placed in the Chalk River reactor in order to get the desired activity [206].

Appendix C
Grimmett's Suggested References on Cobalt-60

The following is an excerpt from Grimmett's letter to Jasper Richardson, April 26, 1951:

> In response to an earlier request of yours for some references touching on our Cobalt-60 program, you will find the following of some help:-

C.W. WILSON, 'Radium Therapy', Chapman &Hall, London, 1945 (this excellent little book will give you a general orientation as regards gamma-ray therapy of cancer.)

COLIEZ, Robert, Journal de radiologie et d'Electrologie XXX, p. 518, 1949 (An early paper outlying the possibilities of Co-60.)

MORTON, and MYERS, Am. J. Roent., Vol. 60, p. 816, Dec '48 (new ideas about Co-60 as a substitute for *Ra needles*).

H.F. FREUNDLICH, Acta Rad., XXXIV, p. 115, Jul-Aug '50 (Review of some gamma-active isotopes suitable for therapy, with details of an Iridium Irradiator.)

H. MILLER, Brit. J. Rad., XXIII, p. 731, Dec '50 (2-Mev X-Ray Generator.-Techniques for measuring electronic build-up in water, etc. useful model for our experiments.)

MAYNEORD, Supplement No.11, Brit. J. rad., 1950(I believe I already mentioned this work to you? It is an excellent summary, but don't take his pessimistic remarks about large Co-60 sources too seriously!)

C.A.P. WOOD, and J.W. BOAG, 'Researches on the radiotherapy of Oral Cancer', Report No. 267, H.M. Stationary Office, London, 1950(Most of the physics in this is my work, although you wouldn't think so from the scanty acknowledgements! You will find some ideas here on measurements in general which can be applied to our cobalt program.)

J. R. GREENING, Brit. J. Rad., XXIV, p.204, April '51 'Effective' wavelength in irradiated water.

> This reading will keep you busy for a bit. It is not exhaustive or comprehensive, but I haven't the leisure just now to look out all the pertinent references. I will do so at the earliest opportunity, however. [135]

Perhaps, it was not intentionable but Grimmett still seemed to be upset by what had happened in 1944 at the Radiotherapeutic Research Unit at Hammersmith Hospital with Jack Boag and Paul Howard Flanders. The "scanty" acknowledgement to him in the reference above is very limited, and if it was mainly his work, he had reason to be upset.

Fig. C.1 Grimmett's notebook [140]

He may still have been outdone with Mayneord over the dispute concerning the output of the radium units just prior to the war, telling Richardson not to take too seriously his pessimistic views about the availability of large Co-60 sources. In fact, Freundlich makes a similar comment in his paper, which Grimmett does not mention, and which led Freundlich and the Cambridge group to go with Iridium-191 instead. With the comparatively small neutron capture, cross section of cobalt-59 and the long life of cobalt-60 and the relatively low neutron flux in the British reactor iridium seemed a better choice. It was therefore reasonable to think that large cobalt-60 sources would not be available. The exterior and flysheet of Grimmett's notebook are shown in Figs. C.1 and C.2, respectively.

Appendix C: Grimmett's Suggested References on Cobalt-60

Fig. C.2 Fly sheet of Grimmett's notebook [141]

References

1. Leonard Grimmett. portrait. 1949. Historical Resources Center Image Collection, Historical Resources Center, Research Medical Library, The University of Texas M. D. Anderson Cancer Center. Illustration used by permission.
2. Going critical: 50 years of nuclear energy. *Journal of Nuclear Medicine* 33(12):16N, 29N. Dec 1992
3. Read, J. 1951. Obituary L.G. Grimmett. *British Journal of Radiology* 24(285):522.
4. Grimmett, L.G. 1948. Letter of resignation to Dr. Auger. August 21, 1948. UNESCO archives Paris.
5. Logo of the United States Atomic Energy Commission. 1946–1974. *United States Atomic Energy Commission.*
6. Baker Estate. 1950. Historical resources center image collection, Historical Resources Center, Research Medical Library, The University of Texas M. D. Anderson Cancer Center. Illustration used by permission.
7. Texas State Hospital Underway. 1942. *JAMA* 120(4):298.
8. Bates, W.B. 1956. History and development of the Texas medical center. *Texas Gulf Coast Historical Society.*
9. Fletcher, G.H. 1979. The early days at M. D. Anderson. *The Cancer Bulletin* 31(2):S8.
10. *The Starkville (Miss.) News*, Friday, April 25, 1947, Society Section.
11. *The Starkville (Miss.) News*, Friday, May 2, 1947, Society Section.
12. Macon, N.D. 1976. *Clark and the Anderson: A personal profile*, 188. Houston: The Texas Medical Center Houston.
13. Mayneord, W.V., and J. Cipriani. 1947. The absorption of gamma-rays from Co-60. *Canadian Journal of Research* 25(6):303.
14. Excerpt from Dr. Leonard's Grimmett Diary, 1948, Leonard Grimmett papers, 1923, 1945–1950, 1964, Historical Resources Center, The University of Texas M.D. Anderson Cancer Center
15. *The Starkville (Miss.) News*, Friday, February 20, 1948, Society Section.
16. Cobalt for cancer tested. *The Science News-Letter* 53(18):275. 1948.
17. Medical and Scientific Director of the American Cancer Society, Inc. Memo to the Divisions of the American Cancer Society, Inc. re: Radioactive Cobalt. April 30, 1948. Directors office records: Ernst Bertner and R. Lee Clark, 1941–1975, The President's office records, 1941–1975, Historical Resources Center, Research Medical Library, The University of Texas M. D. Anderson Cancer Center.
18. Fletcher, G.H. Memo to R. Lee Clark re: Radioactive Cobalt. [ca. 1948]. Directors office records: Ernst Bertner and R. Lee Clark, 1941–1975, The President's office records, 1941–1975, Historical Resources Center, Research Medical Library, The University of Texas M. D. Anderson Cancer Center.

19. Clark, R.L. Letter to Shields Warren. February 5, 1949. Directors office records: Ernst Bertner and R. Lee Clark, 1941–1975, The President's office records, 1941–1975, Historical Resources Center, Research Medical Library, The University of Texas M. D. Anderson Cancer Center.
20. Clark, R.L. 1949. Letter to Albert Thomas. May 18, 1949. Directors office records: Ernst Bertner and R. Lee Clark, 1941–1975, The President's office records, 1941–1975, Historical Resources Center, Research Medical Library, The University of Texas M. D. Anderson Cancer Center.
21. Cancer Work in Peril City May Lose Atomic Center. 1949. *Headline—The Houston Post*.
22. Atomic Cancer Treatment. 1949. *Editorials—The Houston Post*.
23. *The first twenty years of the University of Texas M. D. Anderson Hospital and Tumor Institute*, 62. The University of Texas M.D. Anderson Hospital and Tumor Institute, Houston, Texas, 1964.
24. Macon, N.D. 1976. *Clark and the Anderson: A personal profile*, 189. Houston, TX: The Texas Medical Center Houston.
25. Grimmett, L.G. Letter to Norah Grimmett. February 8, 1949. Leonard Grimmett papers, 1923, 1945–1950, 1964, Historical Resources Center, The University of Texas M.D. Anderson Cancer Center
26. Grimmett, L,G. Letter to Richardson. September 23, 1921. University of Texas Archives & Institute of Physics. Richardson Archives.
27. Grimmett, L.G. Letter to Richardson. July 11, 1927. University of Texas Archives & Institute of Physics. Richardson Archives.
28. Grimmett LG. Letter to Richardson. December 22, 1929. University of Texas Archives & Institute of Physics. Richardson Archives.
29. Richardson, O.W., and L.G. Grimmett. 1930. The emission of electrons under the influence of chemical action at lower gas pressures. *Proceedings of the Royal Society of London. Series A, Containing Papers of a Mathematical and Physical Character*. 130(812):217–238.
30. Grimmett, L.G. Letter to Richardson. May 18,1932. University of Texas Archives & Institute of Physics. Richardson Archives.
31. Read, J. 1951. Obituary notices. *The Proceedings of the Physical Society* 64(Section A): 1143–1144.
32. Flint, H.T., and L.G. Grimmett. 1930. Measurement of the distribution of gamma rays around a four-gram mass of Radium. *British Medical Journal* 2(3628):98–99.
33. Flint, H.T., L.G. Grimmett, E.R. Carling, and S. Cade. 1934. Radium teletherapy: Latest modification of the westminster apparatus and its use. *The British Medical Journal*. 1(3823):653–655.
34. Spear, F.G., and L.G. Grimmett. 1933. The biological response to gamma rays of Radium as a function of the intensity of radiation. *British Journal of Radiology* 6(67): 387 (1 July 1933).
35. The Empire Cancer Campaign. *The Times of London*, April 11, 1934.
36. Spear, F.G., and L.G. Grimmett. 1935. Supplementary note on "the biological response to gamma rays of radium as a function of the intensity of radiation". *British Journal of Radiology* 8(88):231–234 (1 Apr 1935).
37. Grimmett, L.G. Letter to Richardson. May 20, 1932. University of Texas Archives & Institute of Physics. Richardson Archives.
38. Grimmett, L.G. 1932. A direct-reading γ-ray electroscope. *Proceedings of the Physical Society* 44(4):445–450.
39. Grimmett, L.G. 1933. A sensitivity-control for the Lindemann electrometer. *Proceedings of the Physical Society* 45(1):117–119.
40. Wellcome library archives in London. Dr. Constance Annie Wood archives.
41. Annual Report to the Board of Regents and the Chancellor of the University of Texas. 1951. Houston: University of Texas M.D. Anderson Hospital for Cancer Research.
42. Email from Professor Anders Brahme of Karolinska Institute to Peter Almond. 2004.

References

43. Wood, C.A.P., L.G. Grimmett, and T.A. Green. 1938. Medical Research Council—special report series, no. 231. Report on radium beam therapy research 1934–1937, 69. Grimmett's copy from MDACC archives.
44. Sievert, R.M. 1933. The new apparatus for teleradium treatment used at radiumhemmet. *Acta Radiologica* 14:197–206.
45. Grimmett, L.G. 1937. A five-gramme radium unit, with pneumatic transference of radium. *British Journal of Radiology* 10(110):105–117 (1 Feb 1937).
46. Wood, C.A.P., L.G. Grimmett, and T.A. Green. 1938. Medical Research Council—special report series, no. 231. Report on radium beam therapy research 1934–1937, 73. Grimmett's copy from MD Anderson archives.
47. Wilson, C.W. 1948. *Radium therapy. Its physical aspects*, 168. London: Chapman and Hall.
48. Grimmett, L.G., and J. Read. 1935. An application of a new dense tungsten alloy in teleradium therapy. *British Journal of Radiology* 8(94):661–665 (1 Oct 1935).
49. Grimmett, L.G., and J. Read. 1936. Protection problems associated with the use of a 5-GM. Radium unit. *British Journal of Radiology* 9(107):712–731 (1 Nov 1936).
50. Wilson, C.W. 1948. *Radium therapy. Its physical aspects*, 71. London: Chapman and Hall.
51. Grimmett, L.G. Postcard to Norah Grimmett. August 31, 1937. Leonard Grimmett papers, 1923, 1945–1950, 1964, Historical Resources Center, The University of Texas M.D. Anderson Cancer Center
52. Grimmett, L.G. Postcard to Norah Grimmett. September 13, 1937. Leonard Grimmett papers, 1923, 1945–1950, 1964, Historical Resources Center, The University of Texas M.D. Anderson Cancer Center
53. Eve, A.S., and L.G. Grimmett. 1937. Radium beam therapy and high-voltage X-rays. *Nature* 139(3506): 52–55.
54. Grimmett, L.G. 1938. Nuclear physics and medicine. *Nature* 141(3564):311–314.
55. Wood, C.A.P., L.G. Grimmett, and T.A. Green. Medical Research Council—special report series, no. 231. Report on radium beam therapy research 1934–1937. 1938. 70. Grimmett's copy from the MDACC archives. Illustration used by permission.
56. Letter from D. W. Kerst to Grimmett on May 13, 1942 from University of Illinois University Archives.
57. Letter from Grimmett to Kerst. 1942. University of Illinois University Archives.
58. History of Hospital Physicists' Association 1943–1983: Hospital Physicists' Association. 1983. 6–9.
59. Grimmett, N. Letter to Leonard Grimmett. February 9, 1949. Leonard Grimmett papers, 1923, 1945–1950, 1964, Historical Resources Center, The University of Texas M.D. Anderson Cancer Center
60. Grimmett, L.G. Letter to Norah Grimmett. February 10, 1949. Leonard Grimmett papers, 1923, 1945–1950, 1964, Historical Resources Center, The University of Texas M.D. Anderson Cancer Center
61. Grimmett, L.G. Letter to Sir Owen Richardson. June 29, 1943. University of Texas Archives—The Institute of physics. Richardson Collection.
62. Declaration Attested by Supervising Teacher, 1943. Signed by Grimmett and Richardson. From University of Texas Archives—The Institute of physics. Richardson Collection.
63. History of Hospital Physicists' Association 1943–1983: Hospital Physicists' Association. 1983. 101.
64. Grimmett's Transcript. King's College University of London Archives.
65. U.K. National Archives. Ref.: FD113386. Kew, England.
66. Shalek RJ. Personal Communication.
67. Grimmett, N. Letter to Leonard Grimmett. [ca. February 13, 1949]. Leonard Grimmett papers, 1923, 1945–1950, 1964, Historical Resources Center, The University of Texas M.D. Anderson Cancer Center

68. Joseph Neeham, F.R.S. Farewell Dinner Menu, n.d., Leonard Grimmett papers, 1923, 1945–1950, 1964, Historical Resources Center, The University of Texas M.D. Anderson Cancer Center
69. Excerpt from Dr. Leonard's Grimmett Diary, 1947, Leonard Grimmett papers, 1923, 1945–1950, 1964, Historical Resources Center, The University of Texas M.D. Anderson Cancer Center.
70. "Pavlovaelev, Radiologfru" newsclipping, Dagens Nyheter, July 16, 1948, Leonard Grimmett papers, 1923, 1945–1950, 1964, Historical Resources Center, The University of Texas M.D. Anderson Cancer Center.
71. "To Dr. L.E. Grimmett From the National Sciences Department. UNESCO." pastel print, undated, Leonard Grimmett papers, 1923, 1945–1950, 1964, Historical Resources Center, The University of Texas M.D. Anderson Cancer Center. Illustration used by permission.
72. Fermi, E., E. Amaldi, O. D'Agostino, F. Rasetti, and E. Segre. 1934. Artificial radioactivity produced by neutron bombardment. *Proceedings of the Royal Society of London. Series A, Containing Papers of a Mathematical and Physical Character* 146(857):483–500.
73. McLennan, J.C., L.G. Grimmett, and J. Read. 1935a. Artificial radioactivity produced by neutrons. *Nature* 135(3404):147.
74. McLennan, J.C., L.G. Grimmett, and J. Read. 1935b. Production of radioactivity by neutrons. *Nature* 135:505.
75. Lawrence, E.O. 1934. Radioactive Sodium produced by deuton bombardment. *Physical Review* 46(8):746.
76. Brucer, M. 1990. *A Chronology of Nuclear Medicine*, 292.
77. Rotblat, J. 1935. Induced radioactivity of Nickel and Cobalt. *Nature* 136:515.
78. Sampson, M.B., L.N. Ridenour, and W. Bleakney. 1936. The isotopes of Cobalt and their radioactivity. *Physical Review* 50(4):382.
79. Risser, J.R. 1937. Neutron-induced radioactivity of long life in Cobalt. *Physical Review* 52(8):768–772.
80. Livingood, J.J., and G.T. Seaborg. 1941. Radioactive isotopes of Cobalt. *Physical Review* 60(12):913.
81. Nelson, M.E., M.L. Pool, and J.D. Kurbatov. 1942. The characteristic radiations of Co-60. *Physical Review* 62(1–2):1–3.
82. Mitchell, J.S. 1946. Applications of recent advances in nuclear physics to medicine. *British Journal of Radiology* 19(228):481–487.
83. Grimmett LG. Letter to Norah Grimmett. February 9, 1949. Leonard Grimmett papers, 1923, 1945–1950, 1964, Historical Resources Center, The University of Texas M.D. Anderson Cancer Center
84. *The first twenty years of the University of Texas M. D. Anderson Hospital and Tumor Institute*, 24–27, 53. The University bof Texas M.D. Anderson Hospital and Tumor Institute, Houston, Texas, 1964. Illustration used by permission.
85. Notes on Grimmett from discussion with Harriet Awapara. Private communication.
86. Grimmett, L.G. Letter to Norah Grimmett. February 23, 1949. Leonard Grimmett papers, 1923, 1945–1950, 1964, Historical Resources Center, The University of Texas M.D. Anderson Cancer Center
87. Grimmett, L.G. Letter to Norah Grimmett. March 16, 1949. Leonard Grimmett papers, 1923, 1945–1950, 1964, Historical Resources Center, The University of Texas M.D. Anderson Cancer Center
88. Grimmett, L.G. Letter to Norah Grimmett. March 8, 1949. Leonard Grimmett papers, 1923, 1945–1950, 1964, Historical Resources Center, The University of Texas M.D. Anderson Cancer Center
89. Almond P. Personal Communication.
90. Grimmett, L.G. Memo entitled "Provisional 1949 Work Plan for Physics Section", February 15, 1949, Research, radiotherapy, radiation physics, proposed program, 1946–1951, Directors office records: Ernst Bertner and R. Lee Clark, 1941–1975, The

References

President's office records, 1941–1975, Historical Resources Center, Research Medical Library, The University of Texas M. D. Anderson Cancer Center.

91. Grimmett, L.G. Memo to Clark re: additional help for physics workshop on January 23,1950, Physics Department, 1949–1956, Directors office records: Ernst Bertner and R. Lee Clark, 1941–1975, The President's office records, 1941–1975, Historical Resources Center, Research Medical Library, The University of Texas M. D. Anderson Cancer Center.
92. Physics staff department. photograph. 1950. Historical Resources Center Image Collection, Historical Resources Center, Research Medical Library, The University of Texas M. D. Anderson Cancer Center; 1950. Illustration used by permission.
93. Cartoon from the Houston Post. April 2, 1949. Illustration used by permission.
94. Grimmett, L.G. Memo entitled "Proposals for the use of Cobalt-60 in Radiotherapy", August 12, 1949, Research, radiotherapy, radiation physics, proposed programs, 1949, Directors office records: Ernst Bertner and R. Lee Clark, 1941–1975, The President's office records, 1941–1975, Historical Resources Center, Research Medical Library, The University of Texas M. D. Anderson Cancer Center.
95. Painter, T.S. Letter to William G. Pollard. June 23, 1949. Directors office records: Ernst Bertner and R. Lee Clark, 1941–1975, The President's office records, 1941–1975, Historical Resources Center, Research Medical Library, The University of Texas M. D. Anderson Cancer Center.
96. Brucer, M. Letter to Clark. July 6, 1949. Grimmett, Leonard G. (Ph.D.) physicist, John P. McGovern Historical Collections and Research Center, Houston Academy of Medicine-Texas Medical Center Library.
97. Brucer, M. Letter to Leonard Grimmett. August 12, 1949. Directors office records: Ernst Bertner and R. Lee Clark, 1941–1975, The President's office records, 1941–1975, Historical Resources Center, Research Medical Library, The University of Texas M. D. Anderson Cancer Center.
98. Lough, S.A. Letter to R. Lee Clark. August 18, 1949, Research, radiotherapy, radiation physics, proposed programs, 1949, Directors office records: Ernst Bertner and R. Lee Clark, 1941–1975, The President's office records, 1941–1975, Historical Resources Center, Research Medical Library, The University of Texas M. D. Anderson Cancer Center.
99. Grimmett, L.G. 1949. Composite of Grimmett's design of Cobalt-60 unit, research, radiotherapy, radiation physics, proposed program, Directors office records: Ernst Bertner and R. Lee Clark, 1941–1975, The President's office records, 1941–1975, Historical Resources Center, Research Medical Library, The University of Texas M. D. Anderson Cancer Center. Illustration used by permission.
100. Grimmett, L.G. Notes, "A Cobalt-60 Irradiator for Cancer Treatment", February 4, 1950, Research, radiotherapy, radiation physics, proposed program, 1949, Directors office records: Ernst Bertner and R. Lee Clark, 1941–1975, The President's office records, 1941–1975, Historical Resources Center, Research Medical Library, The University of Texas M. D. Anderson Cancer Center.
101. Grimmett, L.G. Grimmett's drawing of a loading Cobalt-60 machine, research, radiotherapy, radiation physics, proposed program, 1949, Directors office records: Ernst Bertner and R. Lee Clark, 1941–1975, The President's office records, 1941–1975, Historical Resources Center, Research Medical Library, The University of Texas M. D. Anderson Cancer Center. Illustration used by permission.
102. Grimmett, L.G., H.D. Kerman, M. Brucer, G.H. Fletcher, and J.E. Richardson. 1952. Design and construction of multicurie Cobalt teletherapy unit. A preliminary report. *Radiology* 59:19–29.
103. Clark, R.L. Letter to Marshall Brucer. September 21, 1949, Archives, Research—radiotherapy—Cobalt 60 project—MDA and Oak Ridge, John P. McGovern Historical Collections and Research Center, Houston Academy of Medicine-Texas Medical Center Library

104. Brucer, M. Letter to R. Lee Clark. November 8, 1949. Directors Office Records: Ernst Bertner and R. Lee Clark, 1941–1975, The President's Office Records, 1941–1975, Historical Resources Center, Research Medical Library, The University of Texas M. D. Anderson Cancer Center.
105. Clark, R.L. Letter to Marshall Brucer. November 18, 1949. Directors office records: Ernst Bertner and R. Lee Clark, 1941–1975, The President's office records, 1941–1975, Historical Resources Center, Research Medical Library, The University of Texas M. D. Anderson Cancer Center.
106. Brucer, M. Letter to A. H. Holland, Jr. December 2, 1949. Archives, Research—radiotherapy—Cobalt 60 project—MDA and Oak Ridge, John P. McGovern Historical Collections and Research Center, Houston Academy of Medicine-Texas Medical Center Library
107. Brucer, M. Letter to Paul C. Aebersold. December 2, 1949. Research—radiotherapy—Cobalt 60 project—MDA and Oak Ridge, John P. McGovern Historical Collections and Research Center, Houston Academy of Medicine-Texas Medical Center Library
108. Brucer, M. Letter to Leonard Grimmett. December 2, 1949. Directors office records: Ernst Bertner and R. Lee Clark, 1941–1975, The President's office records, 1941–1975, Historical Resources Center, Research Medical Library, The University of Texas M. D. Anderson Cancer Center.
109. Report of meeting held December 20, 1949, Concerning muticurie Cobalt 60 sources. December 20, 1949. Directors office records: Ernst Bertner and R. Lee Clark, 1941–1975, The President's office records, 1941–1975, Historical Resources Center, Research Medical Library, The University of Texas M. D. Anderson Cancer Center.
110. Joint proposal to the Atomic Energy Commission by the Medical Division of the Oak Ridge Institute of Nuclear Studies and the University of Texas Postgraduate School of Medicine M. D. Anderson Hospital for Cancer Research. 1950. Directors office records: Ernst Bertner and R. Lee Clark, 1941–1975, The President's office records, 1941–1975, Historical Resources Center, Research Medical Library, The University of Texas M. D. Anderson Cancer Center.
111. Grimmett, L.G. Memo to Clark re: estimated cost of materials required for the fabrication of the head of the Cobalt-60 irradiator (excluding suspension mechanism and control panel) on January 25, 1950. MDACC Archives.
112. *The first twenty years of the University of Texas M. D. Anderson Hospital and Tumor Institute*, 24. The University of Texas M.D. Anderson Hospital and Tumor Institute, Houston, Texas, 1964.
113. Fletcher, G., and L.G. Grimmett. Report on a mission to Washington, D.C. to attend a meeting on Cobalt-60, called by the Atomic Energy Commission. February 12–14, 1950. Directors office records: Ernst Bertner and R. Lee Clark, 1941–1975, The President's office records, 1941–1975, Historical Resources Center, Research Medical Library, The University of Texas M. D. Anderson Cancer Center.
114. Attendants Roster for a meeting on multicurie Cobalt 60 teletherapy sources. February 13, 1950. Directors office records: Ernst Bertner and R. Lee Clark, 1941–1975, The President's office records, 1941–1975, Historical Resources Center, Research Medical Library, The University of Texas M. D. Anderson Cancer Center.
115. Grimmett, L.G. Notes on a Meeting Called by the U.S. Atomic Energy Commission, Oak Ridge National Laboratory for the discussion of a Cobalt-60 irradiator. February 13, 1950. Directors office records: Ernst Bertner and R. Lee Clark, 1941–1975, The President's office records, 1941–1975, Historical Resources Center, Research Medical Library, The University of Texas M. D. Anderson Cancer Center.
116. Grimmett, L.G. Effects of Washington meeting on the design for Cobalt-60 unit. 1950. Directors office records: Ernst Bertner and R. Lee Clark, 1941–1975, The President's office records, 1941–1975, Historical Resources Center, Research Medical Library, The University of Texas M. D. Anderson Cancer Center.

References

117. Macon, N.D. 1976. *Clark and the Anderson: A personal profile*, 195. Houston, TX: The Texas Medical Center Houston.
118. Grimmett, L.G. 1950. The use of Cobalt-60 in medicine. Directors office records: Ernst Bertner and R. Lee Clark, 1941–1975, The President's office records, 1941–1975, Historical Resources Center, Research Medical Library, The University of Texas M. D. Anderson Cancer Center.
119. Photo of mockup. 1950. MDACC Archives.
120. Grimmett, L.G. 1950. 1000-Curie Cobalt-60 irradiator. *Texas Reports on Biology and Medicine* Winter Issue (Oct–Dec).
121. Grimmett, L.G., and G.H. Fletcher. 1953. A 1000-Curie Cobalt-60 irradiator. *Acta Union Internationale Contre Le Cancer* IX(1):43–47.
122. Grimmett, L.G. Letter to W. Binks. August 9, 1949. Radiation Physics Departmental Archives, The University of Texs M. D. Anderson Cancer Center
123. Contract agreement between Oak Ridge Institute of Nuclear Studies and the University of Texas. July 1950. Directors office records: Ernst Bertner and R. Lee Clark, 1941–1975, The President's office records, 1941–1975, Historical Resources Center, Research Medical Library, The University of Texas M. D. Anderson Cancer Center.
124. *The first twenty years of the University of Texas M. D. Anderson Hospital and Tumor Institute*, 66. The University of Texas M.D. Anderson Hospital and Tumor Institute, Houston, Texas, 1964.
125. *The first twenty years of the University of Texas M. D. Anderson Hospital and Tumor Institute*, 67. The University of Texas M.D. Anderson Hospital and Tumor Institute, Houston, Texas, 1964.
126. The University of Texas M.D. Anderson groundbreaking. photograph. 1950. Historical Resources Center Image Collection, Historical Resources Center, Research Medical Library, The University of Texas M. D. Anderson Cancer Center. Illustration used by permission.
127. General Electronic X-Ray Corporation. Letter to Leonard Grimmett. March 27, 1951. Directors office records: Ernst Bertner and R. Lee Clark, 1941–1975, The President's office records, 1941–1975, Historical Resources Center, Research Medical Library, The University of Texas M. D. Anderson Cancer Center.
128. Grimmett, L.G. Report on mission to Milwaukee, Chicago and Binghamton. 4–9 April 1951. Directors office records: Ernst Bertner and R. Lee Clark, 1941–1975, The President's office records, 1941–1975, Historical Resources Center, Research Medical Library, The University of Texas M. D. Anderson Cancer Center 1951.
129. Figure 7. Photograph of the Control Cabinet. Oak Ridge Institute of Nuclear Studies. Quarterly Report (July 1–Sept 30, 1953). 1953.
130. Kerman, H. 1994. Personal communication. 25th Annual radiation oncology clinical research seminar. Early history of the development of Cobalt teletherapy units.
131. Freundlich, H.F., J.L. Haybittle, and R.S. Quick. 1950. Radio-Iridium teletherapy. *Acta Radiologica* 34:115–134.
132. Freundlich, H.F., and J.L. Haybittle. 1953. An improved Iridium-192 teletherapy unit. *Acta Radiologica* 39:231–241.
133. Grimmett, L.G. Letter to Dr. Jasper Richardson on January 25, 1951. From MDACC Radiation Physics Departmental Archives.
134. Grimmett, L.G. Letter to Dr. Jasper Richardson on March 19, 1951. From MDACC Radiation Physics Departmental Archives.
135. Grimmett, L.G. Letter to Dr. Jasper Richardson on April 26, 1951. From MDACC Radiation Physics Departmental Archives.
136. Cobalt 60 therapy.*Newsweek*, May 28, 1951.
137. Doctor Grimmett, Cancer expert, dies suddenly. *Houston Chronicle*, May 28, 1951.

138. Trout, E.D. Letter to R. Lee Clark. June 20, 1951. Directors office records: Ernst Bertner and R. Lee Clark, 1941–1975, The President's office records, 1941–1975, Historical Resources Center, Research Medical Library, The University of Texas M. D. Anderson Cancer Center.
139. Cobalt-60 unit. photograph. undated. Historical Resources Center Image Collection, Historical Resources Center, Research Medical Library, The University of Texas M. D. Anderson Cancer Center.
140. Grimmett's notebook cover. Peter R. Almond private collection.
141. Grimmett's notebook, flysheet. Peter R. Almond private collection.
142. Grimmett, L.G. Memo entitled "Proposals for the Educational Programme of the Physics Department". August 9, 1949. Research, radiotherapy, radiation physics, proposed program, 1949, Directors office records: Ernst Bertner and R. Lee Clark, 1941–1975, The President's office records, 1941–1975, Historical Resources Center, Research Medical Library, The University of Texas M. D. Anderson Cancer Center.
143. Grimmett, L.G. Memo to Clark. June 7, 1950. Physics Department, 1949–1956, Directors office records: Ernst Bertner and R. Lee Clark, 1941–1975, The President's office records, 1941–1975, Historical Resources Center, Research Medical Library, The University of Texas M. D. Anderson Cancer Center.
144. Grimmett, L.G. Memo entitled "Proposals for Construction of an Automatic Dose Contour Plotting Machine", August 12, 1949, Research, radiotherapy, radiation physics, proposed program, 1949, Directors office records: Ernst Bertner and R. Lee Clark, 1941–1975, The President's office records, 1941–1975, Historical Resources Center, Research Medical Library, The University of Texas M. D. Anderson Cancer Center.
145. Grimmett, L.G. Memo to Clark re: Mr. Rasmussen of Kelley-Koett Mfg. Co., September 21, 1949, Research, radiotherapy, radiation physics, proposed program, 1949, Directors office records: Ernst Bertner and R. Lee Clark, 1941–1975, The President's office records, 1941–1975, Historical Resources Center, Research Medical Library, The University of Texas M. D. Anderson Cancer Center.
146. Clark, R.L. Letter to Mr. Rasmussen, September 28, 1949, Research, radiotherapy, radiation physics, proposed program, 1949, Directors office records: Ernst Bertner and R. Lee Clark, 1941–1975, The President's office records, 1941–1975, Historical Resources Center, Research Medical Library, The University of Texas M. D. Anderson Cancer Center.
147. Grimmett, L.G. Memo to Clark re: Physicist-technician for isotope work, May 22, 1951, Physics Department, 1949–1956, Directors Office Records: Ernst Bertner and R. Lee Clark, 1941–1975, The President's Office Records, 1941–1975, Historical Resources Center, Research Medical Library, The University of Texas M. D. Anderson Cancer Center.
148. Grimmett, L.G. Memo to Clark in December 1950 and reply from Clark to Grimmett from MDACC archives radiotherapy—radiation physics (proposed programs)—research 1950–1951; 1948–1949, John P. McGovern Historical Collections and Research Center, Houston Academy of Medicine-Texas Medical Center Library
149. Grimmett, L.G. 1950. Report on some measurements made on a westinghouse "Autoflex" diagnostic set. From MDACC Radiation Physics Department Archives.
150. Scintillation detector and ovoids. photograph. undated. Historical Resources Center Image Collection, Historical Resources Center, Research Medical Library, The University of Texas M. D. Anderson Cancer Center. Illustration used by permission.
151. Grimmett, L.G. Memo to Clark re: money for developing a wavelength independent film, July 6, 1950, Research, radiotherapy, radiation physics, proposed program, 1946–1951, Directors office records: Ernst Bertner and R. Lee Clark, 1941–1975, The President's office records, 1941–1975, Historical Resources Center, Research Medical Library, The University of Texas M. D. Anderson Cancer Center.

152. Fricke, H., and S. Morse. 1927. The chemical action of Roentgen rays on dilute ferrous sulfate solutions as a measure of dose. *American Journal of Roentgenology* 18:430–432.
153. Shalek, R.J., and C.E. Smith. 1969. Chemical dosimetry for the measurement of high energy photons and electrons. *Annals of the New York Academy of Sciences* 161(1):44–62.
154. Oddie, T.H. 1950. Results of uptake and excretion tests with radio-iodine. *British Journal of Radiology* 23(270):348–354.
155. van Roojen, J. 1937. Radiating surfaces. *British Journal of Radiology* 10(117): 650–676 (1 Sept 1937).
156. Grimmett, L.G. Memo to Clark re: Substandard weekly calibration of the x-ray machines, June 21, 1949, Radiation therapy, Department 1948–1956, John P. McGovern Historical Collections and Research Center, Houston Academy of Medicine-Texas Medical Center Library.
157. Grimmett, L.G. Letter to Scheer of Instrumentfirma Gustaf Rose on May 2, 1949. From MDACC Radiation Physics Department Archives.
158. Grimmett, L.G. Letter to Lauriston S. Taylor at the National Bureau of Standards on July 13, 1949. From MDACC Radiation Physics Department Archives.
159. Grimmett, L.G. Letter to Binks at the National Physical Laboratory on August 9, 1949. From MDACC Radiation Physics Department Archives.
160. Notes on Conference with Dr. Fletcher, January 20,1951, Radiation therapy, Department 1948–1956, John P. McGovern Historical Collections and Research Center, Houston Academy of Medicine-Texas Medical Center Library.
161. Fletcher, G. Memo to Grimmett re: Calibration of therapy machines on April 4, 1950. From MDACC Radiation Physics Department Archives.
162. Grimmett, L.G. Memo to Fletcher re: Calibration of X-ray machines on April 5, 1950. From MDACC Radiation Physics Department Archives.
163. Heflebower, R.C. Memo to Fletcher re: responsibilities and roles of radiotherapy and physics, September 15, 1950, Research, radiotherapy, radiation physics, proposed program, 1946–1951, Directors office records: Ernst Bertner and R. Lee Clark, 1941–1975, The President's office records, 1941–1975, Historical Resources Center, Research Medical Library, The University of Texas M. D. Anderson Cancer Center.
164. Fletcher, G. Memo to Grimmett re: Metal Ovoids, April 26, 1951, Radiology Department (diagnostic radiology and radiotherapy), 1948–1956 6, Directors office records: Ernst Bertner and R. Lee Clark, 1941–1975, The President's office records, 1941–1975, Historical Resources Center, Research Medical Library, The University of Texas M. D. Anderson Cancer Center.
165. Fletcher, G. Memo to Grimmett re: Treatment Chair and Adjustment of Light Localizer, May 2, 1951, Radiology Department (diagnostic radiology and radiotherapy), 1948–1956 6, Directors office records: Ernst Bertner and R. Lee Clark, 1941–1975, The President's office records, 1941–1975, Historical Resources Center, Research Medical Library, The University of Texas M. D. Anderson Cancer Center.
166. Fletcher, G. Memo to Grimmett re: Justification of the use of the share of the Institutional Grant of the American Cancer Society given to combined projects—radiology and physics, May 3, 1951, Radiology Department (diagnostic radiology and radiotherapy), 1948–1956, Directors office records: Ernst Bertner and R. Lee Clark, 1941–1975, The President's office records, 1941–1975, Historical Resources Center, Research Medical Library, The University of Texas M. D. Anderson Cancer Center.
167. Fletcher, G. Memo to Heflebower re: Checking of the light localizer of maxitron 250, May 7, 1951, Research, radiotherapy, radiation physics, proposed program, 1949, Directors office records: Ernst Bertner and R. Lee Clark, 1941–1975, The President's office records, 1941–1975, Historical Resources Center, Research Medical Library, The University of Texas M. D. Anderson Cancer Center.

168. Grimmett, L.G. Handwritten note to Mrs. Rita Hendley, 1951, Radiology Department (diagnostic radiology and radiotherapy), 1948–1956, Directors office records: Ernst Bertner and R. Lee Clark, 1941–1975, The President's Office Records, 1941–1975, Historical Resources Center, Research Medical Library, The University of Texas M. D. Anderson Cancer Center.
169. Fletcher, G. Memo to Grimmett re: Weekly calibration of the Maxitron, May 14, 1951, Radiology Department (diagnostic radiology and radiotherapy), 1948–1956, Directors office records: Ernst Bertner and R. Lee Clark, 1941–1975, The President's office records, 1941–1975, Historical Resources Center, Research Medical Library, The University of Texas M. D. Anderson Cancer Center.
170. Fletcher, G. Memo to Clark re: Maxitron, May 15, 1951, Radiology Department (diagnostic radiology and radiotherapy), 1948–1956 6, Directors office records: Ernst Bertner and R. Lee Clark, 1941–1975, The President's office records, 1941–1975, Historical Resources Center, Research Medical Library, The University of Texas M. D. Anderson Cancer Center.
171. Shepley, T. Private Communication.
172. Grimmett, L.G. Memo to Fletcher re: 250 KV Maxitron, May 15, 1951, Radiology Department (diagnostic radiology and radiotherapy), 1948–1956, Directors office records: Ernst Bertner and R. Lee Clark, 1941–1975, The President's office records, 1941–1975, Historical Resources Center, Research Medical Library, The University of Texas M. D. Anderson Cancer Center.
173. Fletcher, G. Memo to Clark re: Chronological order of events on Maxitron Light Localizer, May 15, 1951, Radiology Department (diagnostic radiology and radiotherapy), 1948–1956, Directors office records: Ernst Bertner and R. Lee Clark, 1941–1975, The President's office records, 1941–1975, Historical Resources Center, Research Medical Library, The University of Texas M. D. Anderson Cancer Center.
174. Fletcher, G. Memo to Clark re: Standard procedures in the new therapy machines, May 16, 1951, Radiology Department (diagnostic radiology and radiotherapy), 1948–1956, Directors office records: Ernst Bertner and R. Lee Clark, 1941–1975, The President's office records, 1941–1975, Historical Resources Center, Research Medical Library, The University of Texas M. D. Anderson Cancer Center.
175. Welman, G.A. Letter to Fletcher, May 16, 1951, Radiation therapy, Department 1948–1956, John P. McGovern Historical Collections and Research Center, Houston Academy of Medicine-Texas Medical Center Library.
176. Fletcher, G. Memo to Clark re: 250 KV Maxitron, May 21, 1951, Radiology Department (diagnostic radiology and radiotherapy), 1948–1956, Directors office records: Ernst Bertner and R. Lee Clark, 1941–1975, The President's office records, 1941–1975, Historical Resources Center, Research Medical Library, The University of Texas M. D. Anderson Cancer Center.
177. Grimmett, L.G., G.H. Fletcher, and E.B. Moore. 1954. An improved light localizer for X-ray therapy. *Radiology* 62(4):589–593.
178. Grigg, E.R.N. 1965. *The trail of the invisible light*, 302. Springfield, IL: Charles C Thomas
179. Cormack, D., and P. Munro. 1999. Cobalt-60: A Canadian perspective part 1: The development of kilocurie sources. *Canadian Medical Physics Newsletter* 45(1):10–14.
180. Ibid. 1999. Part 2: The saskatoon story. *Canadian Medical Physics Newsletter* 45(2): 38–41.
181. Munro, P. 1999. Cobalt-60: A Canadian perspective part 3: London, Ontario and the "peacetime bomb". *Canadian Medical Physics Newsletter* 45(3):64–69.
182. Litt, P. 2000. *Isotopes and innovation MDS Nordion's first fifty years, 1946–1996*. Montreal: McGill-Queens University Press.
183. Neil, R.H., W.E. Costolow, and O.N. Meland. 1953. Design and construction of a simple applicator for 1000 Curies of Cobalt 60. *Radiology* 61:408–410.
184. Grigg, E.R.N. 1965. *The trail of the invisible light*, 307–308. Springfield: Charles C Thomas.
185. Heppenheimer, T.A. 2006. How to detect an atomic bomb. *American Heritage of Invention and Technology* 21:44–51.

186. The Lewis L Strauss Page. http://www.smokershistory.com/Strauss.htm.
187. Livingood, J.J., and G.T. Seaborg. 1938. Long-lived radio Cobalt isotopes. Letters to the editor. *Physical Review* 53(10):847–848.
188. P.R. Almond's personal recollection of Fletcher's remarks at his retirement dinner 1984.
189. Johns, H.E., L.M. Bates, E.R. Epp, et al. 1951. 1,000-Curie Cobalt-60 units for radiation therapy. *Nature* 168(4285):1035–1036.
190. Kerman, H.D, and J.T. Ling. 1955. Supervoltage radiation therapy: The University of Lousiville-Louisville General Hospital Cobalt-60 unit. *The Journal of the Kentucky State Medical Association* 1061–1065.
191. Grigg, E.R.N. 1965. *The trail of the invisible light*, 307. Springfield, IL:Charles C Thomas.
192. Grigg, E.R.N. 1965. *The trail of the invisible light*, 313. Springfield, IL:Charles C Thomas.
193. Fletcher, G.H. 1980. *Textbook of radiotherapy*, 3rd ed. Philadelphia: Lea and Febiger.
194. Chairman of the Medical Division, Oak Ridge Institute of Nuclear Studies Under contract with the United States Atomic Energy Commission. 1954. A standard Cobalt 60 teletherapy source capsule. *British Journal of Radiology* 27(319):410–412.
195. Cohen, L., T.E. Schultheiss, and R.C. Kennaugh. 1995. A radiation overdose incident: Initial data. *International Journal of Radiation Oncology Biology Physics* 33(1):217–224.
196. Ministerio de Energia y Minas. Comsion Nacional de Seguridad Nuclear y Salvaguardias. Accidente de contaminacion con 60Co. CNSN-IT-001. Mexico 1984.
197. Dr. Leonard Grimmett's Diary, 1947, Leonard Grimmett papers, 1923, 1945–1950, 1944, Historical Resources Center, The University of Texas M.D. Anderson Cancer Center.
198. United Nations Scientific Committee on the Effects if Atomic Radiation (UNSCEAR). Survey of Medical Usage and Exposures. Radiotherapy Equipment worldwide (1991–1996).
199. Private Communication from P. Grochowska, International Atomic Energy Agency, Vienna. Directory of Radiotherapy Centers (DIRAC), 2012.
200. Morton, J.L., A.C. Barnes, G.W. Callendine Jr, and W.G. Myers. 1951. Individualized interstitial irradiation of cancer of the uterine cervix using Cobalt 60 in needles, inserted through a lucite template; a progress report. *American Journal of Roentgenology and Radium Therapy* 65(5):737–748.
201. Yordy, J.S., P.R. Almond, and L. Delclos. 2012. Development of the M. D. Anderson Cancer Center gynecologic applicators for the treatment of cervical cancer: historical analysis. *International Journal of Radiation Oncology Biology Physics* 82(4):1445–1453 (Mar 15).
202. Almond, P.R., P.J. Biggs, B.M. Coursey, et al. 1999. AAPM's TG-51 protocol for clinical reference dosimetry of high-energy photon and electron beams. *Medical Physics* 26(9):1847–1870.
203. Roberts, J.E. 1999. *Meandering in medical physics: A personal account of hospital physics*, 16. Bristol, Pa.: Institute of Physics Pub.
204. Physicist Grimmett is successful musician also. *Houston Chronicle*, 1951.
205. "A Private Recital of Piano Music Presented by Leonard G. Grimmett", January 30, 1951. Leonard Grimmett papers, 1923, 1945–1950, 1964, Historical Resources Center, The University of Texas M.D. Anderson Cancer Center; 1951.
206. Glasstone, S. 1956. Sourcebook on Atomic Energy. St. Martin's Street, London: Macmillan and Co. Limited.

Index

A
Accredited Dosimetry Calibration Laboratories (ADCL), 119
Aebersold, Paul, 52, 56
AEC, 44, 56, 57, 59, 61
American Cancer Society, 7, 86, 91
Atomic bomb, 26, 106, 131
Atomic Center, 5, 8, 31, 47
Atomic Energy of Canada Limited (AECL), 110, 132
Auger, Pierre, 31, 32
Awapara, Jorge, 43

B
Baker estate, 57
Baker, Captain James, 10, 42, 43, 46, 49
Baylor Medical College, 74
Bertner, Ernst W., 1, 3
Best theratronics, 110
Betatron, Allis Chalmners, 17, 24, 76
Binks, W., 63, 87
Boag, Jack, 92
Bonner, Tom, 44–46, 74
Braestrup, Carl, 88
Bragg–Gray cavity theory, 118
Bragg, William, 22
British Institute of Radiology (BIR), 5, 6, 18, 24, 31
British Journal of Radiology, 18, 39, 97, 107, 112, 123
Brucer, Marshal, 38
Bryant Symons & Co, 106

C
Cervical cancer, 5
Cervical cancer applicators, 86, 87
Cesium-137 (Cs^{137}), 106, 118
Chalk river, 5, 59, 63, 72
Chemical dosimetry, 81, 82
Chicago Tumor Institute, 69, 105
Cipriani, A. J., 5, 38, 64, 97, 108, 110
Clark, R. Lee, 44–49, 52, 56, 72, 74–76, 87, 90, 92
Clinical research, 17
Cobalt-60 (Co^{60}), 37, 53, 77
Cole, A., 76
Conference of Allied Ministers of Education, 30, 31
Contamination, 101, 102, 113
Costolow, William, 105
Critz, Mary Walker, 4, 6
Cutler, Max, 69

D
Damon Runyon Memorial Fund for Cancer Research, 58
Diagnostic radiology, 77, 126
Dosimeters, 78–80

E
Education, 11, 12, 30, 31, 74
Eldorado, 59, 60, 63, 67, 68, 110
Eldorado mining and refining company, 63, 104, 108, 110

E (*cont.*)
Equipment, 22, 26, 29, 31, 75, 89
Errington, R. F., 104, 108, 110
Eve, Arthur Stewart, 36, 76

F
Fifth International Cancer Congress Paris 1950, 63, 108
Film dosimetry, 80
Fletcher, Gilbert, 43, 46, 47, 80, 84, 85–95
Flint, H. T., 15
Free-air standard ionization chamber, 118, 119
Freundlich, H., 67, 106, 117, 136

G
General electric, 57, 60, 64, 75, 114
Gill, Norah Anastasia, 12
Ground breaking, 3, 57

H
Half-life, 35–38, 62, 67, 102, 106, 112, 127
Hammersmith Hospital, 10, 23–25, 27, 92
Head and neck cancer, 5
Heflebower, Roy, 56, 80, 83
Hevimet, 58, 90–92
High voltage engineering corporation, 10, 114
Hospital Physicists' Association (HPA), 24, 29, 31, 49, 122
Houston, 1, 4–6, 8, 10, 24, 29, 32, 35, 36, 41, 43–47, 49, 51, 56–58, 64, 69, 70, 72, 74, 78, 81, 88
Houston Chronicle, 49, 70
Houston Post, 8, 49
Howard-Flanders, Paul, 25
Howe, Clifton D., 72
Huxley, Julian, 44

I
Imperial College, London, 23
International Atomic Energy Agency (IAEA), 109, 115, 116
Inverse square law, 36, 105
Ionization chambers, 78, 86
Iridium-192 (Ir^{192}), 67

J
Johns, Harold, 60, 63, 67, 104, 109

K
Kelley-Koett Manufacturing Company, 61, 75
Kerman, H., 64, 69, 108
Kerst, Donald, 19
King's College London, 11, 12, 25, 30, 44
Kipling street, 43, 45, 72
Kocian, Trudy, 72, 76
Korean war, 49, 58, 61, 64, 97

L
Lederman, Manuel, 5, 60
Linear accelerators, 31, 114–118
London, Ontario, 68, 104, 108, 110
Los Angeles Tumor Institute, 105, 108, 109
Lough, Dr., 52, 58, 59, 104

M
M D Anderson Foundation, 1, 2, 8, 42, 49
M D Anderson Hospital (MDAH), 1, 2, 6, 7, 10, 17, 20, 29, 40, 43, 48, 49, 56, 68, 75, 81, 111, 122
Manhattan project, 52, 103
Mayneord, V. W., 5, 38, 68
McCarthy, Glen, 58
McLean, Charles, 48, 88, 89, 91, 92
McLennan, Professor Cunningham, 17
Medical Research Council (MRC), 17, 18, 114, 115
Meland, Orville, 105
Mellanby, Edward, 18, 20, 25–27
Mitchell, J. S., 39, 67, 107
Moore, E. Baily, 46–48, 69, 95

N
National Bureau of Standards (NBS), 86, 118
National Institute of Science and Technology (NIST), 118
National Physical Laboratories (NPL), 87
National Research Council of Canada (NRC), 110
Nature, 20, 36–38
Needham, Joseph, 30
Neil, Russell Hunter, 105, 108, 109
Newsweek, 69–71
Noble prize, 11, 26, 30, 35

O
Oak Ridge Institute of Nuclear Studies (ORINS), 48, 51, 52, 64, 69, 86

Index

Oak Ridge National Laboratory (ORNL), 52, 104
Oaks, The, 1, 2, 10, 42
Ovoids, 80, 86, 89–91

P
Painter, Dr. T., 52, 56, 75
Penumbra, 60, 62, 105, 116, 129
Perspex man, 18, 22
Physics Department, 5, 8, 15, 17, 23, 44, 45, 47, 68, 73–75, 88, 93
Physic's machine shop, 46–48, 73
Picker X-Ray Co., 109
Pneumatic transfer, 20, 22, 106
Priority, 23, 45, 50, 76, 89, 132
Protection, 18, 52, 59, 73, 77, 78, 127

R
Radar, 114, 115
Radiation sickness, 12, 29, 30
Radioactive isotopes, 3, 5, 7, 35, 37–39, 47, 76, 107, 117
Radiobiology, 63, 73, 83
Radioiodine (I^{131}), 47
Radiological Society of North America (RSNA), 90
Radium, 5–7, 15–17, 19, 22, 23, 29, 36–38, 42, 53, 62, 80, 86
Radium beam therapy research, 16–18, 22, 35–37, 39, 123
Radium bomb, 15, 19, 51, 128
Radium Institute, 17, 22, 23
Radium teletherapy, 3, 5, 15, 22, 67, 105, 123
Radiumhemet, 5
Radon, 16, 35, 118, 126
Reactor, 37, 39, 59, 67, 104, 107, 108
Read, John, 13, 18, 44
Rice Institute, 36, 42, 44, 74
Rice University, 45
Richardson, Jasper E., 48, 68, 79
Richardson, Owen, 11, 25, 44
Royal Cancer Hospital, 5, 17, 22, 31, 60

S
Scintillation detectors, 79
Shalek, R. J., 68, 80, 91
Shields Warren, 7, 44
Shivers, Alan, 50
Sievert, Rolf, 5, 16–18, 78

Sixth International Congress of Radiology London 1950, 63
Skin reaction, 85, 111, 128
Smithsonian institute, 109
Sodium-24 (Na^{24}), 35, 37
Source size, 60, 62, 105, 129
Spear, F. G., 15, 16, 18, 52
Specific activity, 38, 58–60, 63, 67, 105, 132
Starkville, Mississippi, 4, 6
Strauss, L., 106, 107

T
Tele-radium, 7, 107
Texas medical center (TMC), 2, 6, 49, 64, 72
Texas reports on biology and medicine, 62, 108
Theratron, 110, 112
Theratronics international limited, 110
Thomas, Albert, 49
Thomas, M. H., 59
Thomson, G. P., 26, 29
Tracerlab, 61
Treatment machine calibrations, 87
Trout, Dale, 60, 62, 70, 72
Trump, John, 10, 114
Tungsten alloy, 19, 58, 64, 68, 99

U
UNESCO, 6, 10, 29, 30, 44
United States Atomic Energy Commission (USAEC), 47, 49
University of Louisville, 64, 108
UNSCEAR, 116

V
Van de Graff, 10, 23
Van dr Graff accelerators, 10, 114
Varian associates, 115
Victoreen condenser chambers, 87

W
Westminster hospital, 12, 16–18, 53
Wilson, H. A., 44
Winchell, Walter, 58, 70
Wood, Constance, 18, 20, 22, 23, 25, 85
Wootton, Peter, 88

The manufacturer's authorised representative in the EU is Springer Nature Customer Service Centre GmbH, Europaplatz 3, 69115 Heidelberg, Germany. If you have any concerns regarding our products, please contact ProductSafety@springernature.com

Printed and bound by CPI Group (UK) Ltd, Croydon, CR0 4YY

23/03/2026

02076380-0011